THE INORGANIC ANALYSIS
OF PETROLEUM

THE INORGANIC ANALYSIS
OF PETROLEUM

by

JAMES W. McCOY

Supervising Chemist, Analytical Laboratory
Standard Oil Company of California
Western Operations, Inc.
Richmond, California

CHEMICAL PUBLISHING COMPANY, INC.

© 1962

CHEMICAL PUBLISHING CO., INC.
New York   N.Y.

PREFACE

Although many papers have been published describing methods for the inorganic analysis of petroleum no book has previously appeared devoted exclusively to this subject. The purpose of this work is to provide a laboratory handbook for industrial analysts of various degrees of professional training covering the determination of those elements commonly occurring in various types of petroleum products. The procedures represent, from the author's point of view, a reasonable compromise among the usual conflicting interests of speed, accuracy, and cost, and emphasize manufacturing rather than research applications.

The usual industrial method is written to apply to specific products wherein the element to be determined is normally present within rather well-defined limits of concentration. These methods are then followed as recipes with more or less satisfactory results. Products for which this type of method is prescribed can often be handled very well by instrumental methods of analysis. In spite of the numerous papers on instrumental analysis, however, and the large volume of advertising of this trend, wet chemical methods still predominate in the daily work of a refinery laboratory, and will no doubt continue to do so.

The methods in this book are written to apply to as great a variety of products as possible. Throughout the book, however, suggestions are made which should enable the user to modify or shorten the procedures for specific purposes. Numerous notes have been included in the procedures, which the author believes will be helpful; the information they contain is more easily expressed in this way than by extensive discussion sections.

v

A separate chapter is devoted to each element, and the chapters are arranged in alphabetical sequence by element except bromine, chlorine, and fluorine which are covered in one chapter under the generic name halogens.

It is appropriate to acknowledge here the generous assistance of a number of persons during the years while this book has been in preparation. I know no more able chemist than Mr. Leo Heller of the California Chemical Company, and to him I express special appreciation for many helpful and instructive discussions during twelve years of association. Without the interest and encouragement of Mr. R. C. Vollmar, Chief Chemist of the Richmond Refinery, this book could not have been produced, and it is a pleasure to acknowledge his cooperation in making its publication possible. The evaluation of the analytical procedures was materially aided by the careful work of a number of my associates in the analytical laboratory at Richmond, notably: Messrs. J. A. Bertera, E. I. Pollard, M. G. Smith, D. L. Spurlock, and W. T. Werner.

Finally, I am indebted to my wife for her assistance with the reading of proof; to Mrs. Nora Hutton, for much of the preliminary stenographic work; and, with special thanks, to Mrs. Rosemarie M. Wildebaur, who so ably and so cheerfully accomplished the tedious task of typing the manuscript.

JAMES W. McCOY

May 1962
Richmond, California

CONTENTS

Chapter   *Page*

THE INORGANIC ANALYSIS
OF PETROLEUM

INTRODUCTION

In the following chapters, methods for determining twenty-four common chemical elements naturally present in or purposely added to petroleum and its products are described in detail. It is presumed that, in general, a petroleum analytical laboratory is concerned with non-routine analyses, and with procedures that are hazardous, or require special equipment; for the most part, the book is concerned with these types of analyses, with emphasis on wet chemical methods.

The methods of the American Society for Testing Materials, issued annually by the D-2 Committee on Petroleum Products and Lubricants, have been standard in the petroleum industry for many years, and the methods given in this book are meant to supplement rather than supplant the ASTM Standards. Widely used ASTM procedures, such as the determinations of tetraethyllead by acid extraction, and of sulfur by lamp, or methods with oxygen bomb, are not considered here, for it can safely be assumed that analysts in the petroleum industry are familiar with these methods and already use them regularly.

In Chapter 1, a short account is given of the usual natural inorganic components of petroleum and a number of additives are briefly described. The inorganic analysis of petroleum does not differ materially from that of any other substance once the solution to be analyzed has been prepared. The choice of the method for decomposing the petroleum sample is of paramount importance, however, and this subject is considered in detail in Chapter 2. In subsequent chapters, methods are given for determining the individual elements in various types of samples. In developing these methods for particular applications, frequent use has been made of standard works on inorganic analysis, especially: *Applied Inorganic Analysis* by Hillebrand, Lundell, Hoffman, and Bright; and Sandell's *Colorimetric Determination of Traces of Metals.*

It is assumed that common laboratory instruments, including electrolytic equipment, pH meters and potentiometers, and spectrophotometers, are available. Less common equipment, such as flame photometers, polarographs, emission spectrographs, and X-ray fluorescent spectrometers, are omitted here, or mentioned but briefly.

The important problem of sampling petroleum is the subject of an ASTM publication, *ASTM Manual on Measurement and Sampling of Petroleum and Petroleum Products.* The usual difficulties of mixing, dirty containers, contaminated sample lines, and possible effects of the material of the sample on the container are always a consideration, and for all these specific recommendations are made in the individual chapters.

Chapter 1

THE INORGANIC COMPONENTS OF PETROLEUM

Much effort has been directed to ascertaining the origin and the conditions of formation of petroleum accumulations. A knowledge of the geological relationships, the depths and ages of pools, and their compositional characteristics is useful in exploring and judging the prospects of undeveloped areas. Several hypotheses have been advanced to account for the evolution of petroleum from deposits of organic material in marine sediments into the final form in which it is found today.[1]

The organic origin of petroleum is indicated by the presence of chlorophyll derivatives, nitrogenous compounds, and optically active substances.[2] Marine sediments contain great quantities of micro-organisms, and their role in the formation of petroleum has been investigated.[3] Bacterial reduction usually stops at the production of carboxyl, hydroxyl, amino, and sulfhydryl groups rather than continuing in complete reduction to hydrocarbon, although methane-producing anaerobic bacteria are common. The most likely sources of petroleum are the fatty oils, however, and these are very resistant to bacterial action.

The development of a general theory for the mechanism of petroleum evolution is hampered by the absence of any deposits of intermediate age. There is, however, considerable evidence of catalytic polymerization at low temperatures, and cracking by acid silicates, clays, and sedimentary silicate rocks. A comprehensive review of this evidence has been made by Brooks.[4] The presence of porphyrins and other

3

heat-sensitive substances in petroleum indicates that extreme temperatures did not prevail during the history of formation; the upper limit may have been 140°F, and there is no evidence that high pressures were a factor in the cracking process.

Crude oils produced from very old sedimentary rocks tend to be light paraffinic products containing a large proportion of hydrocarbons of low molecular weight, low in nitrogen and oxygen. Younger formations are heavy naphthenic oils with very little of light hydrocarbons, and appreciable amounts of nitrogen and oxygen. As asphalt appears to be a primary substance, it is likely that asphaltic crudes were laid down in reservoir rocks without catalytic properties.

Many different metallic elements have been reported in crude petroleum. In a study of the oil-soluble components of 25 different crude samples, Jones and Hardy [5] found the most prevalent to be aluminum, calcium, iron, nickel, silicon, sodium, and vanadium. Of these, sodium and vanadium ordinarily comprise more than ten per cent of the total ash; in only a few instances was vanadium less than one per cent. A number of the crudes contained chromium, copper, lead, magnesium, and manganese to the extent of one-tenth to one per cent of the ash, and isolated samples contained traces of boron, cobalt, molybdenum, platinum, potassium, silver, strontium, and titanium. The oil-soluble forms include porphyrin-metal chelated complexes, and probably other complexes with nitrogen-containing compounds and soaps. As these are effective emulsifying agents, they hold water or brine in extremely stable emulsions. Dodd et al.[6] have investigated the film-forming characteristics of crude oils containing metallic complexes. Metals may occur in the same crude in more than one form; and inorganic forms may be present in highly dispersed suspensions, or in the colloidal state.[7]

The properties and structures of the porphyrin complexes have been studied rather extensively,[8, 9] and Groennings [10] has developed a quantitative method for the determination of porphyrins in petroleum. Beach and Shewmaker,[11] as a result of extraction studies, have established two classes of vanadium compounds in petroleum. Class I comprises those that are extractable with aqueous pyridine; and Class II comprises those that are not extractable. By conducting molecular distillations with castor oil as a carrier they estimated the equivalent atmospheric boiling points of the volatile porphyrins to be in the range of 1085° to 1200°F, with molecular weights in the 543–800 range. From measurements of absorbency they conclude that the difference in the volatile and non-volatile aggregates is probably in the peripheral groups rather than in the nitrogen-vanadium bonds; and that the non-volatile porphyrins have large asphaltic or polymeric side chains which distort the nitrogen-vanadium bonds and produce a non-characteristic spectrum. The vanadium porphyrins have been most thoroughly investigated. Nickel porphyrins have similar properties, and it will be seen later that their respective volatilities are important considerations when determining the concentrations of these elements by chemical methods.

Table 1 shows the analyses of six different random samples of crude oils. These are given to show the rather wide range in concentrations of the common elements, but they are not necessarily typical of the particular crudes. The California sample is from the Midway field, and is of interest as the youngest petroleum formation in the United

TABLE 1:1

ANALYSES OF CRUDE OILS

Element	Source					
	ARABIA	CALIFORNIA	CANADA	COLORADO	SUMATRA	VENEZUELA
	parts per million					
Al	0.40	1.3	0.50	0.50	0.80	0.40
Ca *	3.0	7.0	—	—	5.0	10.00
Cr	0.160	—†	0.040	0.080	—	1.10
Cu	0.330	0.80	0.160	0.320	0.20	1.80
Fe	1.20	18.0	2.20	1.30	1.60	14.0
Ni	5.30	31.0	1.60	0.40	8.0	112.0
Si	7.40	3.0	6.90	5.60	—	10.0
Na	0.70	3.90	13.0	4.10	3.90	28.0
V	13.0	7.0	0.84	1.9	0.18	1290.0
	percent					
Ash	0.004	0.009	0.009	0.009	0.004	0.21
N	0.095	0.420	0.098	0.098	0.110	0.65
S	2.10	0.660	0.740	0.740	0.110	5.50

* Determined spectrographically.
† Dashes indicate element not determined.

States. It is estimated to be between ten and thirteen million years old.[4]

In addition to the metallic elements originally present in crude oil, several others are often introduced in removing it from the ground. Emulsified brine and drilling muds (bentonites weighted with barite or hematite) account for the presence of such salts as aluminum, barium, calcium, iron, magnesium, and sodium, combined with bicarbonate, chloride, silicate, and sulfate. These salts are extractable with hot water if a demulsifier is added. The presence of chlorides is especially objectionable because the hydrolysis of these salts produces hydrochloric acid, which causes extensive and rapid corrosion. The pyrolytic decomposition starts at the rather low temperature of 250°F; this type of corrosion introduces additional metals from condensers and piping. A further source of contamination during refining is from the attrition of various catalysts. Crudes containing aliphatic and naphthenic acids are also very corrosive, and they introduce oil-soluble contaminants into petroleum unless they are neutralized early in the refining. Sour crudes containing hydrogen sulfide and mercaptans are destructive to equipment and require special handling.

Although the total ash of crude oil seldom exceeds .05 per cent, and is often as little as 0.001 per cent, the processing of large amounts of petroleum leads to the rapid accumulation of large concentrations of ash in the residua, and consequent mechanical plugging of furnace tubes and related equipment, unless the oil is desalted beforehand. As mentioned before, the volatility of certain metallic compounds may lead to their presence in distillate fractions, their concentrations increasing with the depth of the cut. Some of these alter the selectivity of cracking catalysts,[12] increasing the formation of coke and hydrogen and decreasing the yield of gasoline. The more common metals of this type are nickel, copper, vanadium, and iron. Others, in particular the alkali metals, decrease catalytic activity by participating in what is commonly called an acid-base reaction with the acidic catalyst (probably more properly classified as ion-exchange). This latter effect is permanent and the alkali and alkaline earth metals are true poisons.

Non-volatile vanadium, which concentrates in the residuum from which heavy fuel oil is produced, is extremely corrosive to refractories used in furnaces.[5] Its most serious destructive effect is upon fire-clay bricks. Vanadium oxides form low-melting eutectics with the clay, producing hard glassy slags which are extremely difficult to remove; alumina and magnesia firebrick are much less affected. Sodium salts have a synergistic effect on the attack by vanadium.

Basic nitrogen compounds are temporary poisons of the catalysts used for cracking, the effect decreasing in the following order, according to Mills et al.: [13] quinaldine, quinoline, pyrrole, piperidine, decylamine, and aniline. The effect of ammonia is not serious. Pyrroles also promote the formation of gum in gasoline, and contribute to the sludging tendency of distillate fuel oils in storage. Deal et al.[14] have summarized the numerous nitrogen compounds that have been found in California petroleums.

Having surveyed briefly the various elements that may be said to occur naturally or adventitiously in petroleum, we shall mention some of those that are purposely added to improve specific properties of petroleum products. Inhibitors are used to prevent oxidation, corrosion of bearings, knocking, formation of gum, burning of valves, ignition at surfaces, rusting, and other undesirable reactions. Additives are available for improving such specific properties as the stability of lubricants under extreme pressure, pour point, viscosity index, heat resistance, film strength, oiliness, spreading quality, and detergency. In addition, formulations are used that, themselves, have specific properties. These include emulsifiers, demulsifiers, metal deactivators, and liquid film lubricants. Finally should be mentioned the paint and ink driers, which are metallic salts of the various grades of naphthenic acids produced in petroleum refining.

The most common detergent additives for lubricating oil are metallic salts of phenates, phosphonates, and sulfonates; the inhibitors include carbamates, thiophosphates, and phenates. Studies by Peri [15] with the electron microscope have shown that the effectiveness of sulfonate and phosphate soaps in reducing organic deposits and lacquers is the result of physical peptization. Lead naphthenates, tricresylphosphate, and aliphatic chlorine compounds are used to improve stability of lubricants used at extreme pressures; hindered phenols, copper organic salts, and various amines are used as inhibitors of corrosion and oxidation; fatty oils and organic phosphate esters improve oiliness and spreading properties; high-molecular-weight polymers improve the viscosity index of lubricating oils; long-chain alkyl phosphates are effective demulsifiers; and aromatic amines deactivate metal contaminants.

Table 2 lists the elements for which methods of de-

TABLE 1:2

OCCURRENCE OF THE ELEMENTS IN PETROLEUM

Element	Present by Chance			Additives			
	Original Crude	Contamination and Corrosion	Catalyst Attrition	Gasoline	Lube Oil	Fuel Oil	Driers
Aluminum	(C)	—	C	—	(I)	—	C
Arsenic	(*)	I	—	—	—	—	—
Barium	(R)	—	—	—	C	I	—
Boron	(R)	—	—	(I)	—	—	—
Bromine	—	—	—	C	—	—	—
Calcium	C	—	—	—	C	I	C
Chlorine	C	C	—	C	C	—	—
Chromium	(C)	C	I	—	—	—	—
Cobalt	(R)	—	I	—	—	—	C
Copper	C	C	C	—	(R)	I	C
Nickel	—	I	C	—	—	—	—
Iron	C	C	—	..	—	—	C
Lead	(I)	C	—	C	C	—	C
Manganese	(I)	—	—	C	—	—	C
Molybdenum	(R)	—	I	—	C	—	(R)
Fluorine	C	C	R	—	—	—	(R)
Nitrogen	C	C	—	C	I	C	—
Phosphorus	—	—	I	I	C	I	—
Selenium	—	—	—	—	I	—	—
Silicon	(C)	C	C	—	(C)	—	—
Sodium	C	C	—	—	—	(C)	—
Sulfur	C	C	—	C	C	C	—
Vanadium	C	C	I	—	—	—	—
Zinc	—	C	—	—	C	—	C

C, commonly; I, occasionally; R, rarely; dash, not significant. Parentheses indicate that element is seldom determined. * Arsenic is present in crude but has seldom been determined.

termination are provided, according to their frequency of occurrence in certain products, and their usual source. Although this table is only a rough and incomplete guide, it may be helpful in handling unknown samples. Many exceptions to this classification occur in research analysis, but the analyst ordinarily has the advantage of direct contact with the source of samples submitted for research and he can thus readily obtain estimates of concentrations, information as to the nature of the sample, possible interfering elements, and so forth. No information is given in Table 2 as to the usual concentrations of the elements; this will be covered in the chapters on the individual elements.

References

1. J. G. McNab, P. V. Smith, Jr., and R. L. Betts, "The Evolution of Petroleum"; *Ind. Eng. Chem.*, 44:2556 (1952).
2. T. S. Oakwood, D. S. Shriver, H. H. Fall, W. J. McAleer, and P. R. Wunz, "The Optical Activity of Petroleum"; *ibid.*, 2568.
3. R. W. Stone and C. E. ZoBell, "Bacterial Aspects of the Origin of Petroleum"; *ibid.*, 2564.
4. B. T. Brooks, "Evidence of Catalytic Action in Petroleum Formation"; *ibid.*, 2570.
5. M. C. K. Jones and R. L. Hardy, "Petroleum Ash Components and Their Effect on Refractories"; *ibid.*, 2615.
6. C. G. Dodd, J. W. Moore, and M. O. Denekas, "Metalliferous Substances Absorbed at Crude Petroleum-Water Interfaces"; *ibid.*, 2585.
7. J. H. Karchmer and E. L. Gunn, "Determination of Trace Metals in Petroleum Fractions"; *Anal. Chem.*, 24:1733 (1952).
8. J. G. Erdman, V. G. Ramsey, N. W. Kalenda, and W. E. Hanson, "Synthesis and Properties of Porphyrin Vanadium Complexes"; *Preprints of General Papers, Division of Petroleum Chemistry, American Chemical Society*, Vol. 1, No. 1, p. 247 (Feb. 1956).
9. J. G. Erdman, J. W. Walter, and W. E. Hanson, "The Stability of Porphyrin Metallo Complexes"; *Preprints of General Papers, Division of Petroleum Chemistry, American Chemical Society*, Vol. 2, No. 1, p. 259 (March 1957).
10. S. Groennings, "Quantitative Determination of the Porphyrin Aggregate in Petroleum"; *Anal. Chem.*, 25:938 (1953).
11. L. K. Beach and J. E. Shewmaker, "The Nature of Vanadium in Petroleum"; *Ind. Eng. Chem.*, 49:1157 (1957).
12. G. A. Mills, "Aging of Cracking Catalysts"; *Ind. Eng. Chem.*, 42:182 (1950).
13. G. A. Mills, E. R. Boedeker, and A. G. Oblad, "Chemical Characterization of Catalysts. I. Poisoning of Cracking Catalysts by

Nitrogen Compounds and Potassium Ion"; *J. Am. Chem. Soc.*, 72:1554 (1950).

14. V. Z. DEAL, F. T. WEISS, and T. T. WHITE, "Determination of Basic Nitrogen in Oils"; *Anal. Chem.*, 25:426 (1953).

15. J. B. PERI, "An Electron Microscope Study of the Performance of a Detergent Oil"; *SAE Journal*, 62:40 (1954).

Chapter 2

PREPARATION OF SAMPLES FOR INORGANIC ANALYSIS

With few exceptions, before petroleum can be analyzed for its inorganic components the elements to be determined must be separated from a preponderance of organic material. Many methods for accomplishing this are available, and all are covered in detail in subsequent sections. For most of the elements to be considered the analyst may choose the most expedient procedure from several different possibilities. The procedure selected for a particular element depends upon the type of material to be examined, the concentration and volatility of the element in question, the presence or absence of other elements, the accuracy required, and the final method of determination to be used.

In the methods of decomposing petroleum and its products for inorganic analysis, the preliminary treatments are as follows:

1. Direct Ashing
2. Soft Ashing and Wet Oxidation
3. Direct Wet Oxidation
4. Fusion with Pyrosulfate
5. The Oxygen Bomb
6. The Peroxide Bomb
7. Sodium Dehalogenation
8. Extraction Methods
9. Combustion Methods
10. Alkaline Sulfide Treatment

Following the detailed discussion of these treatments, a few direct methods will be described briefly. These are special procedures whereby certain elements can be determined directly without preliminary isolation, such as the use of a combustion tube, emission spectrography, and x-rays.

Table 2:1 summarizes the common procedures that may

TABLE 2:1

METHODS OF PREPARING SAMPLES FOR ANALYSIS

Element	Preparation	Element	Preparation
Aluminum	1,2,3	Lead *	2,3,8,10
Arsenic *	3,8	Manganese *	1,2,3
Barium	1,2,3	Molybdenum	2,3
Boron	5,8	Nickel *	1,2,3
Bromine *	5,6,7,8,9	Nitrogen	3
Calcium	1,2,3,8	Phosphorus *	1,3,6
Chlorine *	5,6,7,8,9	Selenium *	3,5
Chromium	1,2,3	Silicon	2,3
Cobalt *	2,3,4	Sodium *	1
Copper *	2,3,4,8,10	Sulfur *	5,6,9
Fluorine *	6,9	Vanadium *	1,2,3
Iron *	1,2,3,4	Zinc *	2,3,4

* In direct ashing, these elements, if present, may produce a fused ash or attack the dish.

be used to prepare a sample of petroleum for inorganic analysis. These procedures will be considered in detail in the following sections. As some of these possible procedures may not be convenient in particular applications, specific recommendations will be made later.

1. Direct Ashing

Direct ashing is the time-honored procedure for removing organic matter and determining total inorganic components. We are not concerned here with determining percentage of ash as such, however, but only with the removal of the organic matter.

A. DRY ASHING

Reduced to the essentials, the process of dry ashing consists of burning a weighed sample of oil in a tared dish, and igniting the residue in a furnace until carbonaceous material has been oxidized. The quantity of sample to be burned must yield enough ash to permit the determination of the element in question, and the choice of method for the final determination is governed by the amount of ash that can be prepared conveniently. These details are covered in separate chapters on the individual elements.

B. SULFATED ASHING

The basic procedure for direct ashing may be varied for special purposes. The sample may be burned as usual and the residue drenched with sulfuric acid. The excess of the latter is fumed off on a hot plate, and the residue ignited at about 550°C until most of the carbon is oxidized. The residue is then cooled, re-treated with sulfuric acid, and finally ignited for a few minutes at 775°C.[1] The purpose of this procedure is the conversion of the elements to a fairly definite composition for weighing. The second treatment with sulfuric acid is necessary to convert to sulfate any sulfide formed in the first ignition.

C. WET ASHING

Another variation of the direct ashing procedure has been described by Milner et al.,[2] wherein an oil sample in a Vycor beaker is sludged with sulfuric acid. One-half milliliter of sulfuric acid per gram of sample is stirred into the oil; the mixture is first heated from above with an infra-red lamp, and then reduced to coke by heating on a hot plate. Finally the beaker and coke are ignited in a furnace at 475–525°C in a stream of oxygen until the carbon is oxidized.

This procedure was subsequently modified by Gamble and Jones,[3] who omitted the heating lamp and increased the ratio of sulfuric acid to one milliliter per gram of sample, with a final ignition at 1000°F.

Both of these sulfuric acid coking procedures are convenient for gas oils and lighter products, but a 25-gram sample is the practical limit for residual fuel oils and other heavy stocks. Constant attention is necessary in the early stages to prevent spattering, and a long ignition is required to oxidize carbon. Vycor beakers are advisable in sizes suited to the capacity of the furnace available. The beakers are somewhat attacked after being used a few times, and there is evidence of a small loss of the inorganic material by fusion into the bottom of the beaker. This was especially marked in determining vanadium in residual fuel oils by the sulfuric acid coking method.

The basic procedure for direct ashing may be modified also by the addition of capturing agents, notably calcium or zinc oxide in order to prevent the volatilizing of phosphorus. Phosphorus initially present in a variety of forms is completely retained by ashing the sample in the presence of either of these oxides.

Having outlined some of the possible variations in the basic procedure we shall now examine three other variables and consider their influence on the recovery of the elements we have chosen to handle by direct ashing. Referring to Table 2:1 we find these elements to be: aluminum, barium, calcium, chromium, iron, manganese, nickel, phosphorus, sodium, and vanadium. We must, in addition, take account of the possible presence of any of the other elements listed in Table 2:1. The three variables to be considered are the composition of the vessel used for the ashing, the rate of burning, and the temperature of the final ignition.

A. COMPOSITION OF VESSEL

In general, the dish or crucible used is either of glazed porcelain, or of platinum. For unknown samples porcelain should always be used.

Lead, cobalt, copper, vanadium, and manganese may fuse with porcelain glaze to some extent, and cannot be quantitatively removed. Alkali halides also may attack the surface and induce the penetration of other elements.

Platinum is likely to be attacked by arsenic, copper, lead, phosphorus, selenium, and sulfur, and—in the presence of carbonaceous material—by cobalt, nickel, and zinc. When using platinum ware carbonaceous material should be burned off at a low temperature with free access of air to protect the metal from reduction products of ferric oxide, phosphate, and sulfate. Alkali halides attack platinum if heated above 1000°C. Platinum may be used for igniting oxides of aluminum, barium, calcium, chromium, magnesium, manganese, and molybdenum. To summarize: Porcelain should be used for igniting iron, nickel, and phosphorus; platinum, for manganese, sodium, and vanadium; either may be used for aluminum, barium, calcium, and chromium.

B. RATE OF BURNING

The second variable under consideration, the rate of burning, is controlled by limiting the access of air, by limiting the area of the burning surface, and by external heating. In general, samples are ignited and allowed to burn freely without the application of a flame. In burning heavy materials, however, such as tars, asphalts, residua, certain crudes, and fuel oils, the application of a small pilot flame may be required to maintain the temperature at or above the fire point. Light stocks may be burned in high-form crucibles to minimize the area of surface, but the usual container is an evaporating dish.

Gamble and Jones [3] have shown that the rate of burning of gas oils affects the loss of ash; the more rapid the burning, the lower the recovery. Oils with high fire points tend to crawl over the edge of the container while burning, but Morgan and Turner [4] have concluded that this crawling across the flame boundary is a distillation process, and that metallic elements are not lost. Their conclusion was based on experiments with calcium, iron, and sodium; it will be seen later that vanadium and nickel probably would be lost in part.

C. TEMPERATURE OF FINAL IGNITION

For the purpose of determining the percentage of ash in a sample of oil the chosen temperature must be high enough to permit the components to be brought to a fairly definite form, and yield repeatable results. Repeatability is improved by converting the oxides to sulfates, and igniting at an arbitrary temperature. The temperature that is chosen, however, must always be a compromise. Confining the argument to the metals to be recovered by direct ashing, the approximate temperatures of decomposition, or melting points, of their sulfates are shown in Table 2:2.

TABLE 2:2

STABILITY OF METALLIC SULFATES

Metallic Sulfate	Decomposes at °C
Aluminum	770
Barium	1580
Calcium	1450 (m.p.)
Chromium	700
Iron	480
Manganese	850
Nickel	840 *
Sodium	884 (m.p.)
Vanadium .	700

* The work of Milner et al.[2] indicates that the decomposition of nickel sulfate is complete at 750°C.

The ASTM sulfated ash procedure [1] specifies a final ignition temperature of 775°C. At this temperature, aluminum, chromium, iron, and vanadium would be weighed as oxides; and barium, calcium, manganese, nickel, and sodium, as sulfates. The weights of these components would be repeatable under the conditions of the sulfated ash method. As we are not concerned with determining a percentage of ash as such, but require only that all of the metals in question be recovered, we may select a lower temperature for the final ignition.

When using platinum ware the carbon should be burned at a low temperature to minimize the danger of attack on the container. Ignition at low temperature also avoids the formation of difficultly soluble oxides of aluminum, chromium, and iron. On the other hand, the temperature must be high enough to burn the carbon in a reasonable length of time. A suitable compromise, considering all of these factors, is a final ignition temperature of about 550°C.

Following this discussion of the three variations of direct ashing, it is necessary to consider some additional factors which may prevent the complete recovery of the elements to be handled by these procedures.

Conflicting reports have been published on the volatilities of various elements. Experience and examination of published evidence indicate that most of the losses that occur in direct ashing take place during the burning of the sample rather than in the final ignition. Often the investigators are concerned with different types of samples, and they consequently reach different conclusions as to the volatility of certain elements. For example, it has been shown by Beach and Shewmaker [5] that in certain crudes vanadium is combined in two different types of porphyrins, one of which is volatile and the other relatively non-volatile. If a crude oil that contains these two types of vanadium compounds is separated by distillation, and both the distillate and the residuum are tested, the volatile vanadium will be found mainly in the distillate, and the non-volatile compounds in the residuum.

If, then, one portion of the distillate is burned and dry-ashed and a second portion is wet-ashed, the vanadium content of the second portion will be higher than that of the first, and one would correctly conclude that the vanadium was volatile. From two samples of the residuum, tested similarly, it could be concluded—also correctly—that the vanadium was not volatile.

Each conclusion is correct for the oil considered, but it would be erroneous to state categorically that vanadium is volatile or non-volatile. Similarly, if two samples each of the distillate and residuum were reduced respectively by dry ashing and sulfated ashing, the results for vanadium by both procedures would be the same, but the results for the distillate would be too low. As the loss of vanadium in the sample of distillate occurs during the burning it would do no good to sulfate the residue before ignition. (The foregoing discussion applies to nickel as well as to vanadium.)

With the exception of calcium and barium, the metals under discussion form volatile halides, and low results are obtained by both dry and sulfated ashing. This factor must be considered in the inorganic analysis of crude oils. The only oxide that volatilizes is sodium oxide, but the conversion is not rapid at 550°C. Other metallic oxides to which direct ashing cannot be applied are those of arsenic, molybdenum, and zinc; arsenic trioxide is completely lost in burning, and molybdenum and zinc oxides partially sublime during ignition.

Some oils are subject to stratification; therefore a sample

should never be taken from the top of a container without thorough mixing. Crude oils especially are heterogenous and must be homogenized before sampling. Heavy stocks must be heated to reduce viscosity before they can be mixed effectively. External drafts can blow ash out of the dish during burning and ignition, and care must be taken to protect the vessel from this effect. Entrainment of particles of ash results from heat currents rising from the burning surface; this is related to the rate of burning, discussed previously. Finally, turbulence at the surface of the burning sample is a serious factor in causing losses. Turbulence is created by water, or by widely divergent boiling points of components of the oil. Water may be removed from samples with high flash points by heating the weighed sample in a drying oven for several hours. Alcohol is fairly effective for removing water; it is added directly to the weighed sample and then heated on a steam plate. Several hours may be required to remove all the water, and two treatments are occasionally necessary.

In the preceding discussion it has been shown that the nature of the oil sample must be taken into account when selecting the dry or wet modification of the direct ashing

TABLE 2:3

APPLICATION OF WET AND DRY ASHING

Element	Light Distillate	Heavy Distillate	Residuum	Crude Oil
Aluminum	(d)*	d	d	w
Barium	(d)	d	(d)	(d)
Calcium	(d)	d	d	d
Chromium	(d)	(d)	(d)	(w)
Iron	d	w	d	w
Manganese	d	(d)	(d)	(w)
Nickel	(d)	w	d	w
Phosphorus	d†	d†	d†	d†
Sodium	d	d	d	d
Vanadium	(d)	w	w	w

* Parentheses indicate that element is seldom present or determined in that type of product.
† With calcium or zinc oxide as a capturing agent.

procedure. (We shall not have occasion to apply the sulfated ash method.) To summarize the conclusions and provide a guide for handling samples we offer below an arbitrary general classification of characteristic petroleum products to which direct ashing is most frequently applied:

1. Light distillate (gasoline, naphtha, kerosene, stove oil, diesel fuel, etc.)
2. Heavy distillate (gas oils, lube cuts, etc.)
3. Residuum (fuel oil, still bottoms, tar, asphalt)
4. Crude Oil

Table 2:3 shows the dry and wet modifications of the direct ashing procedure applicable to each element, designated d and w respectively.

Any elements present in light distillates are ordinarily non-volatile additives or contaminants.

2. Soft Ashing and Wet Oxidation

In the procedure of soft ashing and wet oxidation, a weighed sample is transferred to a dry 400-ml beaker, the beaker is enclosed in a No. 2 tin can with the top cut out, and the sample is ignited by heating the can with a flame. When the fire burns out the beaker is removed from the can, and loose carbon is burned from the sides by heating with a blast burner, care being taken not to heat the bottom directly. The beaker is then cooled, sulfuric acid is added, and the remaining carbonaceous material is destroyed by addition of nitric acid and other oxidizing agents. (Details of the wet oxidation process are covered in Section 3.)

With the exception of phosphorus, sodium, and vanadium, all of the elements for which direct ashing is applicable can be recovered by this procedure of soft ashing and wet oxidation. For obvious reasons phosphorus cannot be recovered by this method, and there is a possibility of loss of metallic elements in some samples. Sodium is not handled by wet oxidation because there is too much contamination from the beaker and cover glass through the action of hot acid solutions.

Table 2:1 shows that, in addition to the elements discussed in Section 1, cobalt, copper, lead, molybdenum, silicon, and zinc can also be handled by soft ashing and wet oxidation. Special care must be taken, however, in applying this procedure to lead and silicon. If any overheating occurs during the ashing operation lead may fuse into the walls or bottom of the beaker, and it cannot be recovered. If silicon is combined in a silicone it may be volatilized in part during the burning. If lead is present it is sometimes advisable to add a few drops of fuming sulfuric acid to the weighed sample before ignition. This treatment converts lead soaps, etc., to lead sulfate (m.p. 1170°C) which does not fuse under the specified conditions. The sulfuric acid pretreatment is also effective for recovering trace metals from heavy distillates. Because of the importance of accurately determining small amounts of metallic contaminants in gas oils used as feed stocks for catalytic cracking this application is next considered in some detail.

In cracking processes that utilize aluminum silicate catalysts, two important effects arising from metallic contaminants in the feed stock are recognized. The first of these, loss of activity as the result of permanent poisoning by alkali and alkaline earth metals, is the more serious. Nitrogen bases are temporary poisons, and their effect, as well as that of potassium, has been studied by Mills et al.[6] Loss of selectivity, the second effect, is caused by the presence in the feed of elements of the first transition series. As these metals are dehydrogenation catalysts, they cause an increase in formation of coke and hydrogen. Copper and nickel, which are the most effective in this respect, are of about equal activity. Vanadium, chromium, and iron are less active, decreasing in that order. As cobalt and manganese are not present in the distillates used as feed stocks, they need not be considered.

As these contaminants are present in very small concentrations large samples are required to determine them by wet chemical methods. In general, all these elements are volatile to some extent, and precautions must be taken to avoid losses during the burning of the sample. Table 2:4 shows the effect of adding a few drops of fuming sulfuric

TABLE 2:4

EFFECT OF SULFURIC ACID PRETREATMENT ON
RECOVERY OF METALS FROM GAS OILS

Sample	ppm Iron		ppm Nickel		ppm Vanadium	
	Soft Ash	H_2SO_4	Soft Ash	H_2SO_4	Soft Ash	H_2SO_4
I	0.33	0.42	0.15	0.70	1.1	1.9
II	0.94	1.16	Nil	0.15	0.10	0.18
III	0.41	0.66	0.03	0.15	0.40	0.84
IV	2.35	2.90	0.08	0.28	1.30	1.50

acid to the weighed sample before burning. In the columns
headed "*Soft Ash*" are the results obtained by burning a
sample in a beaker and wet-oxidizing the residue. The results in the columns headed H_2SO_4 were obtained in the
same way, but with the sulfuric acid pretreatment. Evidently the sulfuric acid effectively decomposes the porphyrins and soaps with which the metals are combined. The
procedure is much less tedious than the wet-ashing technique
described in Section 1, and gives comparable results.

To indicate the concentration range of the metals that are
active dehydrogenation catalysts, the minimum, maximum,
and average concentrations found in 36 cuts of heavy gas
oil from 8 different crude sources are given in Table 2:5.

TABLE 2:5

CONCENTRATIONS OF TRANSITION ELEMENTS IN
CUTS OF HEAVY GAS OIL

Metal		ppm	
	Minimum	Maximum	Average
Chromium	Nil	0.2	0.03
Copper	Nil	2.3	0.50
Iron	0.20	6.0	2.00
Nickel	Nil	1.0	0.35
Vanadium	Nil	3.5	0.32

As chromium is only about one-tenth and iron about one-
fiftieth as active as nickel, it can be seen from the table that
these two metals are relatively insignificant, and they therefore need not ordinarily be determined. In addition, a large
proportion of the iron may not be in an oil-soluble form,
and, if nitrogen bases are present, as much as half of the

total copper may be found in filterable sludge. If the metals are not in oil-soluble form their effect on the catalyst beds is much less.

The question of whether soluble metals or total metals should be determined is still open. The soluble metals may be presumed to be those that remain after passing the oil sample through a fritted-glass filter crucible. Oil-soluble copper content, in particular, depends upon the length of time the sample stands, and this must be taken into account in interpreting results. In Table 2:6 are summarized the

TABLE 2:6

EFFECT OF FILTRATION ON METAL CONTENT OF GAS OIL

Metal	Unfiltered, ppm	Filtered, ppm
Copper	0.80	0.55
Iron	4.30	2.10
Nickel	0.43	0.40
Vanadium	0.25	0.22

results on filtered and unfiltered portions of a heavy gas oil. The content of the filterable sludge was 445 ppm, of which 0.08 per cent was copper, and 0.43 per cent was iron. Nickel and vanadium were not detected in the sludge.

The use of the sulfuric acid oxidation leads to the formation of several difficultly soluble sulfates which makes the procedure unsuitable for some combinations of elements. The elements that form difficultly soluble sulfates are aluminum, barium, calcium, chromium, iron, and lead. Of these, aluminum and iron can be brought into solution by diluting with water and boiling. Barium, chromium, and lead do not dissolve, nor does calcium if the sulfuric acid has been fumed for some time. It should be realized that insoluble material may enclose elements that would normally be soluble and thus prevent their complete solution.

Most of the remarks in Section 1 concerning losses apply here. The size and shape of the container in this instance, however, minimizes losses by drafts, and there is less danger of overheating. Supporting the beaker in a can has several advantages: The can serves as a muffle, producing even heating; the beaker hangs in the can supported by its lip and top edge (effectively in an air bath), and not exposed to the direct heat of the flame; the can serves as a safety device should the beaker break while the sample is burning.

The sample should be ignited by placing a burner with a luminous flame under the edge of the can. When ignition occurs the flame is removed and the sample is allowed to burn freely. When the flame burns out the can should not be heated further. The beaker is removed and carbon is burned from the sides with a blast burner, taking care not to heat the bottom directly with the flame. The beaker is then cooled and the residue oxidized by the procedure to be considered in Section 3.

3. Direct Wet Oxidation

The usual medium for wet oxidation is concentrated sulfuric acid because of the following properties: It is a relatively non-volatile vehicle; it has a powerful charring and dehydrating effect on organic material; and it can be heated to high temperatures without boiling. Table 2:7 shows the

TABLE 2:7

TEMPERATURE OF FUMING SULFURIC ACID

Degree of Fuming	Qualitative Description	Approximate Temp.	
		°F	°C
Incipient	A few wisps at surface	240	115
Light	Light fuming from entire surface	300	150
Moderate	Beaker filled with fumes	350–400	175–205
Heavy	Dense white rolling fumes	450	235
Boiling	———	550	290

approximate temperatures of various degrees of fuming of sulfuric acid solutions and gives a qualitative comparison which will be found useful later in the discussion.

The usual oxidizing agents are nitric acid, 30% hydrogen peroxide, and perchloric acid. The order and the mode of applying them are varied with different types of samples, and the choice depends upon the elements to be determined. Thus, if manganese is to be determined volumetrically, perchloric acid should not be used to finish the oxidation because manganese is inconveniently oxidized to manganese dioxide. When molybdenum is to be determined volumetrically after zinc reduction, nitric acid must not be left in the solution, for it is reduced to hydroxylamine and other products that reduce permanganate, causing high results. Hydrogen peroxide is destroyed in the zinc column and causes no difficulty. If chromium is to be determined, phosphoric acid should be added before the oxidation to prevent the separation of $Cr_4H_2(SO_4)_7$. Other restrictions and modifications will be noted in discussions of the individual elements.

As direct wet oxidation is extremely tedious when applied to large samples, it should be used only when the element in question is present in sufficient concentration to permit its being determined in no more than 2 or 3 grams of the original sample. Sensitive colorimetric methods are applicable for aluminum, chromium, copper, iron, lead, manganese, molybdenum, nickel, and vanadium. It should be noted that if a gravimetric determination is contemplated at least 3 mg must be weighed to permit an accuracy of 10%. (The average over-all error in two routine weighings on a macrobalance is about 0.3 mg.) Wet oxidation is not convenient for the determination of barium, calcium, or large amounts of aluminum and iron, but it must be used occasionally.

Some details of the basic method of wet oxidation will now be considered. The process is conveniently carried out in a 400-ml beaker equipped with a ribbed cover glass. The sample is weighed and transferred to a clean, unetched beaker, and a 10- to 15-ml portion of concentrated sulfuric acid is added. When continuing the treatment of a sample reduced to a soft ash as described in Section 2, the sulfuric acid should be added, heated to heavy fumes (see Table 2:7), and fumed for 10 to 15 minutes before the addition of any nitric acid. If nitric acid is added at the outset the loose carbon and tarry residue present will cause excessive foaming, and much time may be lost before it can be brought under control. The prolonged fuming decomposes and disperses the carbonaceous material to a considerable extent, and when nitric acid is finally added its effect is much more rapid.

After thorough digestion, the mixture is cooled somewhat and the addition of nitric acid is begun. The acid should be added in 2- or 3-ml portions with intermittent heating, by which process the carbonaceous coating on the walls will be gradually washed into the bulk of the solution in the bottom of the beaker. It should be noted that the deposit of carbon on the walls becomes saturated with sulfuric and nitric acids; therefore the liquid contents of the beaker should not be swirled while the mixture is fuming, or violent spattering will occur when the hot solution in the bottom touches the deposit on the walls. By continually adding small portions of nitric acid the walls will be cleaned by the refluxing acid in a few minutes.

When oxidizing a sample directly, about 5 ml of nitric acid is added at the outset, and any initial reaction should be allowed to subside before heating the mixture. A heavy product dissolved in a lighter solvent (additives, driers, etc.) spatters badly if it is heated with sulfuric acid before nitric acid is added. Material that has been subjected to vigorous conditions during processing (sulfonic acids, sulfonates, residua, cracked tars, etc.) are extremely difficult to oxidize, and usually require prolonged fuming with sulfuric acid to decompose; these products are also apt to foam with nitric acid.

When the initial effects (spattering, foaming, vigorous reaction) have subsided, the temperature of the mixture is raised to the degree of moderate fuming (see Table 2:7), and the remaining organic material is destroyed by adding nitric acid over the lip of the beaker from a dropping pipet. The oxidizing agent should not be added so rapidly that the mixture is unduly cooled, nor should it be added while the mixture is fuming strongly. The former treatment retards the oxidizing action, and the latter is analogous to dropping the acid on a hot stove; it spatters violently and vaporizes immediately without effective oxidizing action.

Ultimately the continued addition of nitric acid will produce a straw-colored solution which upon continued fuming and treating will darken repeatedly. This effect is caused by refluxing acid which washes small amounts of organic material from the cover glass and returns it to the fuming mixture where it chars. A thin, often invisible, oily film is always formed on the cover glass during a wet oxidation and this is very slowly removed by reflux. By replenishing the sulfuric acid and continuing the addition of nitric acid this film could eventually be removed, but an inordinate length of time would be required. It is therefore advisable to add hydrogen peroxide as soon as a relatively clean solution is obtained. The beaker is removed from the heat, the mixture is cooled to incipient or very light fumes, and a 10-ml portion of 30% hydrogen peroxide is added. The beaker is then returned to the heat and the peroxide boiled off. The peroxide distilling out of the hot solution wets the organic material on the cover much more effectively than does nitric acid, and the solution is less likely to darken when brought to fumes again. It may be necessary, however, to repeat the peroxide treatment once or twice.

If there is evidence at this point that organic matter is still present, the mixture is next treated with perchloric acid in the presence of nitric acid as a moderator. The sulfuric acid solution is cooled below the temperature of incipient fuming, and a 5-ml portion of nitric acid is added and mixed. About 2 ml of perchloric acid is next added and the solution heated slowly to the boiling point of perchloric acid (200°C). In the great majority of cases, when the latter has been volatilized the organic material will have been completely destroyed, and it is seldom necessary to repeat the treatment.

Having covered in detail the basic procedure for wet oxidation we will now briefly mention some modifications for specific elements. If silicon is present in a silicone the wet oxidation should be carried out in an Erlenmeyer flask with an excess of nitric acid present from the beginning to minimize loss by volatilization. If phosphorus is to be determined by wet oxidation a Kjeldahl flask is used, in order to provide more condensing surface; phosphoric acid is lost from beakers containing fuming sulfuric acid. Special grades of hydrogen peroxide must be used when determining phosphorus by wet oxidation. It is worth mentioning that most aromatic carboxylic acids and anhydries, such as benzoic, phthalic, isophthalic, and terephthalic acids, as well as the toluic acids, dissolve completely in warm sulfuric acid, and after being heated to fumes they can be oxidized smoothly and rapidly by slow, continuous addition of 30% hydrogen peroxide alone. Some of these form stable nitrates, and their oxidation with nitric acid is slow and tedious. Sulfonic acids and sulfonates also oxidize readily with peroxide, but the heavy oil often associated with them must be oxidized with nitric acid.

For the determination of arsenic Kjeldahl flasks are used in cleaning up extracts obtained as described in Section 8. When samples containing selenium are to be wet-oxidized the oxidation must be carried out in Kjeldahl flasks fitted during the initial treatment with West condensers. Finally should be mentioned the catalytic wet oxidation used in the Kjeldahl method for nitrogen. Among the many modifications of this method that have been described, the catalysts most frequently used are copper, mercury, and selenium. The details of the Kjeldahl method will be covered in the chapter on nitrogen.

Referring to Table 2:1, it is seen that the elements that can be recovered quantitatively by the ordinary wet oxidation procedure are aluminum, barium, calcium, chromium, cobalt, copper, iron, lead, manganese, molybdenum, nickel, silicon, vanadium, and zinc. In addition, with suitable modifications, the basic procedure may be applied to arsenic, phosphorus, selenium, and organically combined silicon. As remarked before, however, this method is not especially convenient to use in determining barium, calcium, and large amounts of aluminum or iron, principally because of the formation of slightly soluble sulfates, mixed crystals, and highly contaminated insoluble material.

For most metallic elements this method of destroying organic material is the least liable to produce losses. One notable exception should be mentioned, however. In determining zinc in lubricating oil additives that contain large amounts of phosphorus and sulfur, and to a lesser extent in zinc naphthenate driers, results by wet oxidation tend to be low. This disadvantage, caused by excessive creeping of these materials, can be avoided by applying the fusion method described next.

4. Fusion with Pyrosulfate

Although fusion with pyrosulfate requires constant attention and careful control, it is a convenient procedure for handling additives and driers containing cobalt, copper, iron, or zinc in the concentration range of 1 to 10%. After the flux is dissolved in water or a suitable acid, all four of these metals can be determined by volumetric methods without further treatment. As mentioned in the preceding section, it is especially advisable to use the fusion method when determining zinc in certain additives and driers, because low and erratic results are sometimes obtained by wet oxidation. Additives containing both calcium and zinc, when wet-oxidized, often produce an insoluble residue of calcium sulfate which does not redissolve upon dilution. The latter tends to enclose zinc mechanically, and low results are obtained unless the calcium sulfate is first brought into solution by additional treatment. Calcium sulfate from a pyrosulfate fusion is readily soluble in water.

Cobalt, and to a lesser extent iron, may attack porcelain glaze under some conditions. To prevent any fusion of these metals with the surface, the porcelain crucible used for the fusion should be coated with a layer of fused potassium pyrosulfate. A No.2 high-form Coors crucible is satisfactory for the operation. Not more than 2 grams—and preferably a 1-gram sample—is weighed and transferred to the prepared crucible, 4 or 5 grams of potassium pyrosulfate is added and mixed with the sample with a small glass rod. The mixture is then covered with a loose layer of pyrosulfate, and the crucible is heated gently with a flame until the sample ignites. When burning ceases the heat is gradually increased until the flux liquefies. As the heat is increased and carbon begins to oxidize, bubbles of carbon dioxide are evolved, and as the flux thickens some spattering will occur. The heating must be carefully controlled and the crucible rotated constantly to minimize this spattering. Eventually the heating drives off all the water, and the melt solidifies.

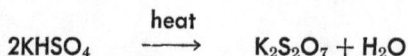

$$2KHSO_4 \xrightarrow{\text{heat}} K_2S_2O_7 + H_2O$$

In some instances the oxidation of carbon is not complete by the time solidification occurs, and it is necessary to cool the crucible, add a few drops of sulfuric acid (not more than 10 drops), and fuse again. If too much sulfuric acid is used the mixture will boil and spatter badly. Precaution should also be taken that none of the flux comes in contact with the tongs holding the edge of the crucible for they are rapidly attacked, introducing chromium, iron, and nickel into the flux. The addition of sulfuric acid should not be omitted in any case because it has the additional effect of volatizing sulfide which is produced by reduction of pyrosulfate, or from sulfur in the sample.

When the operation is complete the melt is cooled and dissolved in water, and the metallic elements may then be determined by appropriate methods to be discussed later. A small amount of acid hastens the solution of the flux, the kind to be used depending upon other considerations.

5. The Oxygen Bomb

The method of determining sulfur and chlorine in petroleum products by use of the oxygen bomb has been an established routine in the petroleum industry for many years. It is applicable to practically any product, and in the case of sulfur, it is virtually free from interferences. As this procedure leaves little to be desired, and as it has been covered in detail elsewhere,[7, 8] its use for the determination of sulfur and chlorine will not be described. We shall discuss here the application of the oxygen bomb to the determination of moderate concentrations of selenium and boron in petroleum products.

Burke[9] has described a satisfactory procedure for the volumetric determination of boron after decomposition in the oxygen bomb. About 10 ml of water containing 1 gram of sodium carbonate is added to the bomb with the fittings prepared in the usual way. Volatile samples may be weighed in gelatine capsules. Oxygen is introduced to a pressure of 27 to 40 atmospheres, depending upon the capacity of the bomb, and the sample is ignited electrically. After ignition the bomb is opened, and the alkaline solution containing the boron as borate is washed out. Boron is then determined alkalimetrically by titration in the presence of mannitol. Details are given in the chapter on boron.

Selenium may be prepared for colorimetric determination by ignition in the oxygen bomb in the presence of dilute sodium hydroxide. The bomb method is best suited for handling from 1 to 5 mg of selenium, determination being made by a suitable colorimetric procedure.

As oxidations in the oxygen bomb are accomplished in a closed system it would be anticipated that there is no possibility of loss until the bomb is opened. The walls of the bomb should be kept polished, however, to avoid retention of the element to be determined, and the bomb must be washed well to ensure quantitative recovery. Excess of oxygen should be released slowly to avoid spray losses while venting the bomb.

Halogens attack the bombs rather severely if more than about 20 mg is present in the charge, and the amount of the sample should be adjusted so that this limit is not exceeded. The maximum amount of sample to be used in an oxygen bomb ignition is 1 gram.

6. The Peroxide Bomb

The Parr peroxide fusion bomb may be used to decompose petroleum products for the determination of some of the non-metallic elements. The Parr Instrument Company publishes a manual [10] in which procedures are described for several elements in a variety of materials. The discussion here will be confined to the decomposition of petroleum samples for the determination of the halogens, phosphorus, and sulfur. It will generally be found, however, that other methods can be applied more conveniently than peroxide fusion. For example, the oxygen bomb, sodium dehalogenation, or extraction by sodium biphenyl is usually preferable for the halogens; direct ashing, for phosphorus; and the oxygen bomb, or direct combustion methods, for sulfur.

Peroxide fusion has a number of disadvantages: Small samples must be used, about 0.3g maximum; large amounts of salts are introduced; high blanks are obtained in the case of phosphorus; nickel and iron are introduced from the fusion cup. Further, samples containing high concentrations of halogen, phosphorus, or sulfur, to which the procedure would normally be applied, are highly reactive, and often ignite the peroxide spontaneously; this procedure is inherently hazardous. In spite of this list of objectionable features, however, it is possible to make safe and accurate determinations by following certain well established practices which will be briefly described. It is recommended that the Parr manual [10] be consulted before any attempt is made to use the peroxide bomb.

The basic procedure involves the intimate mixture of solid sodium peroxide, a combustion aid, an accelerator, and the sample. The mixture is sealed in a bomb and ignited electrically by means of a fuse wire. The sample must not contain water because the heat of the reaction with peroxide may raise the temperature sufficiently to start the oxidation. Volatile liquids may be weighed in gelatine capsules. The capsule is buried in the prepared mixture of the other components, and crushed with a rod. The bomb is then closed and the contents mixed by vigorous shaking. It should be mentioned that many additives containing halogens, phosphorus, and sulfur will ignite the mixture when the capsule is crushed. If the capsule is not crushed, and the sample is left concentrated in one spot, dangerous overheating and extreme pressures are developed upon ignition, and in addition there is likely to be incomplete combustion.

It is advisable to use an accelerator to start the oxidation at a lower temperature and thus ensure complete combustion. Potassium perchlorate is used when determining phosphorus or sulfur, as it produces a smoother, less violent oxidation than potassium chlorate. For determining the halogens, potassium nitrate may be used, but with caution, as it may produce a violent reaction or explosion. Samples that are oxidized with difficulty require a combustion aid to raise the temperature high enough to sustain the reaction. Easily ignited materials with high heats of combustion, such as benzoic acid or sucrose, may be used for this purpose. When using the 22-ml bomb the maximum of combustible material should be 0.5 grams, the maximum of accelerator, 1.0 g (or less, if KNO_3 is used), and a full 15 grams of sodium peroxide. When preparing the charge, the sample, accelerator, and combustion aid are mixed in the fusion cup; the peroxide is added last and mixed thoroughly. If it is necessary to weigh the sample in a capsule, the other components are mixed first, and the capsule is added last and crushed as mentioned previously.

EXTREME CAUTION IS REQUIRED WHEN USING THE PEROXIDE BOMB.

7. Sodium Dehalogenation

Sodium dehalogenation has been applied to a variety of organic materials for reducing organically combined halogens to ionic form for argentometric titrations. The procedure consists of refluxing the sample in a suitable solvent with excess metallic sodium and an alcohol. The method was originated by Stepanow [11] who used absolute ethanol and sodium in specified amounts. Later Drogin and Rosanoff [12] gave empirical rules for calculating the amount of ethanol and sodium to use. Cook and Cook [13] made a critical study of the Stepanow method, modifying the original procedure somewhat, and from time to time other changes have been introduced to make the basic procedure more widely applicable. A standard method [14] is available for the determination of chlorine in lubricating oils and greases containing chlorine additives.

In various modifications of the dehalogenation method several different alcohols have been used, including ethyl, isopropyl, n-butyl, and isoamyl. Different excesses of sodium, and a number of solvents have been specified. Originally some difficulty was experienced in the quantitative recovery of chlorine from nitro-substituted compounds. This was presumably caused by the consumption of sodium in reducing the nitro group to ammonia; a larger excess of sodium is required for such compounds. The procedure is not always satisfactory for volatile substances; and it cannot be used for dehalogenating phenol, or similar materials which themselves consume sodium. Results are low and variable when the method is applied to samples containing less than 10 ppm of halogen. The procedure should be applied with caution to greases and specialty products containing lard oil, substituted phenols, naphthalene, aromatic oils, and various sulfur derivatives, for these may also give erratic results, usually low.

Sodium dehalogenation has not been widely used for the determination of fluorine, and there is little information on this application. Organic fluorine produced from side reactions in the alkylation of hydrofluoric acid, is generally present at low concentrations and the Wickbold combustion method (see Section 9) is more appropriate. Sodium dehalogenation is also satisfactory for the determination of iodine but there is seldom any occasion to make an iodine determination in petroleum products, and this element will not be considered.

The most convenient application of sodium dehalogenation is for the determination of chlorine in viscous materials and polymeric substances with chlorine in the 0.001 to 0.4% range. For such materials benzene is an appropriate solvent. Isopropyl alcohol and about 2 grams of metallic sodium, cut into one-eighth inch cubes, are added and the mixture is refluxed. After refluxing for about an hour the excess of sodium is destroyed by slowly adding methyl alcohol through the top of the reflux condenser. Water is usually specified, but the reaction with methanol proceeds at a more moderate rate, and the solution is blanketed from air by a layer of organic vapors, which minimize the dangers of spattering and fire. The solution is transferred to a beaker, diluted, and acidified with nitric acid.

In general, the two layers need not be separated before potentiometric titration. If organically-combined sulfur is present in the sample, however, it is reduced to sulfide (chlorosulfides are reduced to mercaptans) by sodium, which consumes silver nitrate in the titration.

The presence of sulfur will be apparent from the starting potential. It would be expected that two inflections would be obtained in the titration curve, and that the chloride could be computed from the difference with reasonable accuracy, as in aqueous solutions. In practice, however, where the titration medium is an emulsion of polymeric material, water, and organic solvent, the polymer causes silver sulfide (or mercaptide) formed in the titration to coat the silver electrode. As this coating is viscous and adherent, vigorous stirring is needed to attain equilibrium of the electrode. The time consumed is inordinately great, and the titration "breaks" are indistinct. It is therefore advisable to remove sulfur compounds before starting the titration. To do this the mixture is transferred to a separatory funnel, ethyl ether is added to prevent emulsification, the aqueous layer is drained away, and the organic layer is extracted twice with water. The sulfide is oxidized by boiling with peroxide, and the chloride is then titrated.

8. Extraction Methods

As shown in Table 2:1, the elements that may be separated by various extraction procedures are arsenic, boron, bromine, calcium, chlorine, copper, and lead. These are the elements for which extraction is most convenient; many others could be added to the list, but only these seven will be considered.

The simplest form of extraction is the agitation of an oil sample in a separatory funnel with another liquid which is immiscible with the oil, such as water, various acids, or other aqueous reagents. Separatory methods have been described by Blair,[15] Neilson et al.,[16] and in a manual published by the Petrolite Corporation.[17] All these procedures have been proposed for the extraction of inorganic salts from crude oils. All methods require the dilution of the sample with a suitable solvent, such as xylene, toluene, or benzene (gasoline is not suitable), and agitating the sample with hot water. It has been established, however, that water alone is not sufficient to effect quantitative recovery of salts that are present in brine-and-oil emulsions. These are stabilized by naturally occurring emulsifiers and various destabilizers have been recommended to promote coalescence of the emulsified droplets. The Tret-o-Lites [15] are effective for flocculating emulsified droplets, but some require centrifuging to effect coalescence. Waxy or asphaltic coatings may mechanically enclose salt crystals, and thus protect them from the extraction water. The addition of phenol has been recommended for recovering salts held in this manner.[16] Except for special studies, the chloride ion is the usual component of crude oil determined because its salts tend to hydrolyze upon heating and produce corrosive hydrochloric acid. The effectiveness of desalting equipment is commonly checked by determining the chloride content of crude oil before and after treatment.

Boron gasoline additives are usually water-soluble and can be quantitatively recovered from gasoline by a single water extraction. As boron is present in small concentrations it is most conveniently determined in the water extract by a colorimetric procedure.

As a preliminary treatment for colorimetric determination, copper can be extracted from the raw gasoline obtained from copper sweetening processes by shaking the gasoline with dilute hydrochloric acid. A similar procedure is applicable for the determination of small amounts of calcium in the presence of much zinc in certain additives. Small concentrations of lead can be extracted from distillates with dilute acid, but recovery is incomplete in some instances. The details of all of these procedures will be covered in chapters on the individual elements.

The most widely used extraction procedure in the petroleum industry is that used for the determination of tetraethyl lead in gasoline.[18] The extraction apparatus specified in this method has been applied by Rittershausen and De Gray for the extraction of salts from crude oil [19] and for the extraction of metallic components from various other oils.[20] They found butyl alcohol to be the most effective agent for resolving emulsions. For extracting metallic elements from heavier oils the weighed sample is dissolved in an appropriate thinner and refluxed for 15 to 30 minutes with $6N$ hydrochloric acid. If more than 20 grams of sample is used, bumping may be a problem. In applying hot acid extraction to used oils, Lykken, et al.[21] found that significant amounts of barium, calcium, and silica were left in the oil.

In an extraction procedure widely used for recovering arsenic from straight-run naphthas these are shaken with sodium hypochlorite and 70% sulfuric acid.[22] Maranowski, et al.[23] have reported that recovery of arsenic is incomplete by this method, however, probably because of inadequate contact between the sample and extraction mixture. They have described a wet oxidation procedure which gives better recovery of arsenic.

One of the most convenient and generally applicable methods for recovering organically-combined bromine and chlorine is by the reaction of the sample with disodium biphenyl reagent followed by extraction with nitric acid of the bromide and chloride produced. This method was applied by Pecherer, *et al.*[24] for the determination of bromine and chlorine in gasoline. The procedure and preparation of the reagent was subsequently simplified by Liggett[25] and applied to various organic halogen compounds. The manipulations are simple, volatile materials are easily handled, and the reaction is almost instantaneous. The method is highly recommended for the determination of halogens in petroleum products; the details will be covered in the chapter on the halogens.

To summarize: The various extraction methods are widely applicable and, in many cases, easy to use. Troublesome emulsions and incomplete extractions with some types of samples, however, are often deterrents.

9. Combustion Methods

A. ASTM LAMP

Use of the ASTM lamp for the determination of sulfur in many different petroleum products is a well-established procedure in the petroleum industry. Bromine and chlorine can be quantitatively recovered from some materials, but results are often unsatisfactory. Leaded gasolines, for example, give low results because lead halides condense in the chimney. The lamp method is usually applied when halogens are present in low concentrations, but if the determination is to be finished turbidimetrically concentrations less than 5 ppm give erratic results. The method is not satisfactory for aromatic stocks.

The procedure consists in burning the sample, or a dilution of it, in a wick lamp with air, or in a closed system with a regulated mixture of oxygen and carbon dioxide, the combustion gases being absorbed in a suitable solution, usually dilute sodium hydroxide containing hydrogen peroxide. The method is not used for any determinations covered in this book.

B. WICKBOLD OXYHYDROGEN FLAME

An ingenious oxygen-hydrogen combustion apparatus has been developed by R. Wickbold in Germany, and used by him for the decomposition of large samples of organic materials in the determination of the halogens and sulfur. Two different types of quartz probe burners are available, one for liquids, the other for solids. Wickbold has described the determination of bromine, chlorine, and iodine,[27] and subsequently fluorine,[28] using the burner for solids. Sweetser [29] has also applied this burner for the determination of fluorine in a variety of organic compounds. Later Wickbold developed a suction burner for liquids and described the use of it for determining chlorine and sulfur.[30]

A detailed description of this apparatus and its operation will be found in the Appendix.

10. Alkaline Sulfide Treatment

A solution of ammonium sulfide in alcohol is an especially useful reagent for the recovery of lead and some other heavy metals from petroleum. This procedure has not been previously described in the literature, but it has been found very convenient for the determination of lead in new and used lubricating oils. It has been shown in previous sections that the only satisfactory procedure for removing organic material for the determination of moderate concentrations of lead is wet oxidation. As a two-gram sample is the practical limit for this operation this method, when applied to oils containing 0.1 to 0.5 per cent lead (the usual level of compounding), has the serious disadvantage of yielding a rather small amount of precipitate for weighing.

It has been mentioned that at least 3 mg must be weighed to achieve an accuracy of 10 per cent in a gravimetric determination. In order to obtain a sufficient amount of lead for gravimetric analysis a procedure has been developed for separating the lead from a large sample of oil by precipitation with an alcoholic ammonium sulfide solution. The precipitated lead sulfide is filtered from the oil on paper, the paper and precipitate are oxidized in the wet way, and the lead is determined gravimetrically by one of the conventional methods. The principal difficulty is obtaining a readily filterable lead sulfide precipitate, and many combinations of treatments and solvents were investigated before the final procedure, which is prescribed in the chapters on lead and copper, was adopted.

The most difficult samples to handle are used lubricants from test engines. These oils are often run until they break down completely, so they contain much oxidized oil, varnish, sludge, water, etc., and are seldom completely soluble in any single solvent. For this reason a mixed solvent, containing equal volumes of benzene, acetone, and mixed hexanes (or petroleum ether), has been adopted for the general procedure. The sample is dissolved in the mixed solvent, paper pulp is added, and the mixture is heated with a solution prepared by bubbling hydrogen sulfide into a mixture of alcohol (Formula 30) and ammonium hydroxide. By adding the pulp first, the lead sulfide is precipitated in the interstices of the paper, and filtration is much improved. Precipitation is quantitative with lead, copper, cobalt, and zinc, and probably other elements; iron and manganese are incompletely precipitated.

In concluding this survey of methods to be used for the recovery of inorganic components from petroleum, attention is directed to a study by Barney and Haight [31] of the efficiency of several different methods of recovery as applied to copper, iron, lead, nickel, and vanadium in a fraction of distillate. In the survey are included dry, wet, and sulfated ashing, and extraction with several different solutions.

DIRECT METHODS OF ANALYSIS

Several methods will now be briefly described whereby analyses are made more or less directly on an oil sample in such a way that the destruction of the organic material and the determination of the inorganic component occur essentially simultaneously. This is by no means a complete survey of such methods, and details will not be covered.

11. Combustion Tube

The Dumas method for the determination of total nitrogen is the best established procedure available. The sample is oxidized in a combustion tube by copper oxide in an atmosphere of carbon dioxide, the nitrogen that forms being collected and measured over a strong solution of potassium hydroxide in an azotometer. The procedure gives reliable results in experienced hands provided that certain precautions are taken, the most important of which is the recycling of the evolved gases to ensure the absence of methane and carbon monoxide in the measured nitrogen. Clark [32] points out that this is not an absolute method although it has been called so. He calls attention to the fact that some substances (*e.g.*, derivatives of pyrimidines, purine, and chlorophyll) form non-combustible nitrogeneous *charcoals* which cause low values, and certain compounds cannot be analyzed at all. In addition, it is not applicable for small concentrations of nitrogen, the daily output is small, and extreme care is required in all phases of the determination. Reference should be made to E. P. Clark's book [32] for details of the apparatus and procedure.

The ter Meulen method of determining nitrogen by catalytic hydrogenation,[33] although it was originally proposed in 1924, has been applied extensively only in recent years. As with the Dumas method, experience and attention to detail are required to obtain reliable results, and information is as yet lacking to establish the universality of the method. The procedure consists in pyrolyzing the sample at 900°C in a stream of hydrogen, passing the gases over a nickel-magnesium catalyst at about 300°C. The evolved ammonia is absorbed in boric acid and determined acidimetrically; or, if the amount is small, the ammonia is absorbed in dilute hydrochloric acid and determined colorimetrically. Sulfur poisons the catalyst, and its presence must be noted. Recoveries are low with nitro compounds, but this is not a common form in petroleum. Details of the procedure and an apparatus suitable for multiple determinations are described by Holowchak *et al.*; [34] and Lake [35] has summarized the results of a cooperative testing program in which the Dumas, Kjeldahl, and ter Meulen methods for nitrogen are compared.

The Dietert high-temperature combustion method for sulfur leaves little to be desired as a rapid, accurate, routine procedure, applicable to practically any material that can be weighed in an open boat. The sample is burned in a stream of oxygen at 2400° or 2600°F in the presence of vanadium pentoxide as an accelerator, and alundum as a moderator. The evolved sulfur dioxide is asborbed in an acid solution of potassium iodide containing starch indicator, and the mixture is titrated with standard potassium iodate as the combustion proceeds. The time required for a single determination is usually less than 30 minutes.

This method is applicable to both organic and inorganic materials. With V_2O_5 as a flux, the sulfates of barium, lead, sodium, and other metals are decomposed completely, and the sulfur quantitatively recovered. Halogens and nitrogen sometimes interfere, causing low results, but their effect is not significant in most petroleum samples. Highly chlorinated additives, however, cannot be analyzed for sulfur by this method.

The complete Dietert assembly for determining sulfur is available from the H. W. Dietert Company, Detroit, Michigan, and the manufacturer's instruction manual gives the necessary details for making determinations. The application of the method to petroleum products has been described in an ASTM publication,[36] and Rice-Jones [37] describes its use for determining sulfur in ores and other metallurgical samples.

Combustion tube methods are used for the determination of carbon and hydrogen [32] but these will not be described here. Many modifications of these procedures have been proposed, but they are beyond the scope of this work.

12. The Emission Spectograph

Emission spectroscopy has become a popular analytical tool for the determination of low concentrations of metallic elements in petroleum because of its rapidity and general applicability. It is hardly possible to make an appraisal of the accuracy of the method because of the many variations in technique, but in general routine work the repeatability is of the order of 10 to 15%, and the reproducibility is 20 to 30% for most metallic elements in the range of 1 to 100 ppm. Refinements of technique and special cases often allow higher accuracy.

Spectroscopy can be conveniently applied for the determination of aluminum, barium, calcium, chromium, cobalt, copper, iron, lead, manganese, molybdenum, nickel, silicon, sodium, vanadium, and zinc. The method is not practical for the halogens, nitrogen, and sulfur, nor for volatile elements such as arsenic and selenium. It is not especially sensitive for phosphorus, and there is seldom any occasion to use it for boron, although it is possible.

Karchmer and Gunn [38] have given a summary of direct spectroscopic methods, including techniques with electrode quenching and impregnation, porous cup, and rotating electrode, and Carlson and Gunn [39] have applied the electrode impregnation technique to the direct determination of several common elements in the 1 to 100 ppm range. As the direct method requires special calibration and high sensitivity, routine spectrographic analyses are often made on an ash prepared directly on the matrix (silica, lithium carbonate, etc.). The same considerations of volatility, as mentioned in Section 1 on direct ashing, then apply and some metals may give low results. Gamble and Jones [3] have described a wet-ashing technique whereby the sample is wet-ashed in the presence of magnesium nitrate as a carrier, buffer, and intensifier, with cobalt naphthenate as the internal standard. They report accurate determinations of manganese, nickel, and vanadium in the 0.1–2.0 ppm range, and conclude that similar results could be obtained for copper, iron, and sodium if magnesium nitrate were available sufficiently free from contamination by these elements. Dyroff et al.[40] have compared spectrographic and colorimetric chemical analyses of petroleum ashes; and Calkins and White [41] have applied the electrode quenching technique to the analysis of lubricating oil additives containing barium, calcium, lead, phosphorus, and zinc as well as to blends of these additives.

13. X-ray Methods

The uses of x-rays here may be divided into two general classes, both of which are of the so-called non-destructive type of analysis. The first of these involves absorption of x-rays using an x-ray photometer to measure the absorbency. This method has been applied for the determination of metals in many different petroleum products, and the annual reviews in *Analytical Chemistry* by H. A. Liebhafsky should be consulted for specific references.

Spectrometry by x-ray fluorescence has been used for the determination of tetraethyl lead in gasoline,[42] nickel and vanadium in various petroleum fractions,[43] and barium, calcium, and zinc in lubricating oils; [44] it is particularly useful for the analysis of catalysts. Barieau [45] has applied the absorption edge technique for determining molybdenum and zinc in liquid hydrocarbons.

As these instruments are expensive, use of them is usually limited to research work, and to control work involving a large volume of similar samples. Recent improvements in instruments which allow several different kinds of determinations to be carried out automatically, have considerably enhanced the flexibility of the method.

References

1. Am. Soc. Testing Materials, *1955 Book of ASTM Standards*, Part 5, D874–55:350.
2. O. I. MILNER, J. R. GLASS, J. P. KIRCHNER, and A. N. YURICK, "Determination of Trace Metals in Crudes and Other Petroleum Oils," *Anal. Chem.*, 24:1928 (1952).
3. L. W. GAMBLE and W. H. JONES, "Determination of Trace Metals in Petroleum," *ibid.*, 27:1456 (1955).
4. L. O. MORGAN and S. E. TURNER, "Recovery of Inorganic Ash from Petroleum Oils," *ibid.*, 23:978 (1951).
5. L. K. BEACH and J. E. SHEWMAKER, "The Nature of Vanadium in Petroleum," *Ind. Eng. Chem.*, 49:1157 (1957).
6. G. A. MILLS, E. R. BOEDEKER, and A. G. OBLAD, "Chemical Characterization of Catalysts. I. Poisoning of Cracking Catalysts by

Nitrogen Compounds and Potassium Ion," *J. Am. Chem. Soc.*, 72:1554 (1950).

7. Am. Soc. Testing Materials, Philadelphia, Pa., *1955 Book of ASTM Standards*, Part 5, D129–52:81.

8. *Ibid.*, D808–52T:308.

9. W. M. BURKE, "Boron Determination in Volatile Organic Compounds," *Ind. Eng. Chem., Anal. Ed.*, 13:50 (1941).

10. Parr Manual No. 121, *The Peroxide Bomb Apparatus and Methods*, Parr Instrument Co., Moline, Ill.

11. A. STEPANOW, "Ueber die Halogenbestimmung in organischen Verbindung mittles metallischen Natriums und Aethylalkohol," *Ber.*, 39:4056 (1906).

12. I. DROGIN and M. A. ROSSANOFF, "On the Detection and Determination of Halogens in Organic Compounds," *J. Am. Chem. Soc.*, 38:711 (1916).

13. W. A. COOK and K. H. COOK, "Determination of Nuclear Halogens in Organic Compounds," *Ind. Eng. Chem.*, 5:186 (1933).

14. Am. Soc. Testing Materials, *1955 Book of ASTM Standards*, Part 5, D1317–54T:737.

15. C. M. BLAIR, "Determination of Inorganic Salts in Crude Oils," *Ind. Eng. Chem., Anal. Ed.*, 10:207 (1938).

16. C. A. NEILSON, J. S. HUME, and B. H. LINCOLN, "Determination of Salts in Crude Oil," *ibid.*, 14:464 (1942).

17. Petreco Manual, *Impurities in Petroleum*, Petrolite Corporation, Long Beach, California.

18. Am. Soc. Testing Materials, *1955 Book of ASTM Standards*, Part 5, D526–53T:258.

19. E. P. RITTERSHAUSEN and R. J. DEGRAY, "Determination of Inorganic Salts in Crude Oils," *Ind. Eng. Chem., Anal. Ed.*, 14:947 (1942).

20. E. P. RITTERSHAUSEN and R. J. DEGRAY, "Extraction of Metallic Constituents from Oils," *ibid.*, 14:806 (1942).

21. L. LYKKEN, K. R. FITZSIMMONS, S. A. TIBBETTS, and G. WYLD, "The Determination of Metals in Lubricating Oils," *Petroleum Refiner*, 24:405 (1945).

22. Universal Oil Products Method No. H-296–53.

23. N. C. MARANOWSKI, R. E. SNYDER, and R. O. CLARK, "Determination of Trace Amounts of Arsenic in Petroleum Distillates," *Anal. Chem.*, 29:353 (1957).

24. B. PECHERER, C. M. GAMBRILL, and G. W. WILCOX, "Determination of Bromine and Chlorine in Gasoline," *ibid.*, 22:311 (1950).

25. L. J. LIGGETT, "Determination of Organic Halogen with Sodium Biphenyl Reagent," *ibid.*, 26:748 (1954).

26. Am. Soc. Testing Materials, *1955 Book of ASTM Standards*, Part 5, D1266–55T:679.

27. R. WICKBOLD, "Neue Schnellmethode zur Halogenbestimmung in organischer Substanzen," *Angew. Chem.*, 64:133 (1952).

28. R. WICKBOLD, "Die quantitative Verbrennung Fluor-haltiger organischer Substanzen," *ibid.*, 66:173 (1954).

29. P. B. SWEETSER, "Decomposition of Organic Fluorine Compounds by Wickbold Oxyhydrogen Flame Combustion Method," *Anal. Chem.* 28:1766 (1956).

30. R. WICKBOLD, "Bestimmung von Schwefel- und Chlor-Spuren in organischer Substanzen," *Angew. Chem.*, 69:530 (1957).
31. J. E. BARNEY, II and G. P. HAIGHT, JR., "Efficiency of Recovery Methods," *Anal. Chem.*, 27:1285 (1955).
32. E. P. CLARK, *Semimicro Quantitative Organic Analysis*, Academic Press, Inc., New York (1943).
33. H. TER MEULEN, *Rec. Trav. Chim. Pays-Bas*, 43:1248 (1924).
34. J. HOLOWCHAK, G. E. C. WEAR, and E. L. BALDESCHWIELER, "Application of ter Meulen Nitrogen Method to Petroleum Fractions," *Anal. Chem.*, 24:1954 (1952).
35. G. R. LAKE, "Determination of Nitrogen in Petroleum and Shale Oil," *ibid.*, 24:1806 (1952).
36. Am. Soc. Testing Materials, *ASTM Standards on Petroleum Products and Lubricants*, App. II:944 (1956).
37. W. G. RICE-JONES, "Sulfur in Ores, Concentrates, and Other Metallurgical Samples," *Anal. Chem.*, 25:1383 (1953).
38. J. H. KARCHMER and E. L. GUNN, "Determination of Trace Metals in Petroleum Fractions," *ibid.*, 24:1733 (1952).
39. M. J. CARLSON and E. L. GUNN, "Determination of Trace Metallic Components in Petroleum Oils," *ibid.*, 22:1118 (1950).
40. G. V. DYROFF, J. HANSEN, and C. R. HODGKINS, "Comparison of Spectrochemical and Semimicromethods in Analysis of Petroleum Ashes," *ibid.*, 25:1898 (1953).
41. L. E. CALKINS and M. M. WHITE, "Analyze Additive Lubricants in Minutes Instead of Hours with Spectrographic Method," *Nat'l Petroleum News*, 38, No. 27:519 (1946).
42. F. W. LAMB, L. M. NIEBYLSKI, and E. W. KIEFER, "Determination of Tetraethyllead in Gasoline by X-Ray Fluorescence," *Anal. Chem.*, 27:129 (1953).
43. E. N. DAVIS and B. C. HOECK, "X-Ray Spectrographic Method for the Determination of Vanadium and Nickel in Residual Fuels and Charging Stock," *ibid.*, 27:1880 (1955).
44. E. N. DAVIS and R. A. VAN NORDSTRAND, "Determination of Barium, Calcium, and Zinc in Lubricating Oils," *ibid.*, 26:973 (1954).
45. R. E. BARJEAU, "X-Ray Absorption Edge Spectrometry as an Analytical Tool," *ibid.*, 29:348 (1957).

Chapter 3

ALUMINUM

Although aluminum and its compounds are used extensively in petroleum refining, the occasions for actually determining the element in the analytical laboratory are relatively infrequent. The concentration of aluminum in crude oils rarely exceeds one part per million although it is one of the more prevalent oil-soluble components. A notable exception is Venezuelan crude, in some samples of which the concentration may reach 150 ppm. Aluminum occurs in most residual fuel oils either by concentration, or through contamination by wind-blown dust or by catalyst fines during refining; its concentration is usually in the range of one to ten ppm.

Various adsorptive aluminum clays such as attapulgite, Fuller's earth, Filtrol alumina, and the like are used in petroleum refining for decolorizing, bleaching, or purifying oils, and defective filters may lead to contamination of products by these agents. It is usually sufficient, however, to ash the suspected sample, and determine the percentage of residue, rather than actually to determine aluminum itself. The amount of ash found, and its appearance are ordinarily all that is necessary to establish whether contamination from this source has occurred.

Aluminum may be introduced into lubricating oil by engine wear, either emulsified in the oil, or as an oil-soluble component. It may be necessary to determine it in lubrication research work with test engines. Aluminum naphthenate, prepared from naphthenic acids and aluminum sulfate, is used as an additive in a few special machine oils; as a rule, only the additive is analyzed for aluminum content. A few aluminum salts of fatty acids are used in greases and waterproofing compounds, notably aluminum stearate and palmitate; metallic aluminum pigments are used in formulating paints and roof coatings; bauxite is used as a neutralizer and desiccant in the Hydrofluoric Acid Alkylation process. Aluminum oxide or silicate is commonly used as a vehicle for hydroforming, reforming, cracking, and platforming catalysts, and their attrition may lead to the presence of aluminum in the products of these catalytic processes. Little difficulty is ever caused by catalyst attrition, however, because of the insolubility of the fines in petroleum products; the contaminating substance drops rapidly to the bottom of storage tanks.

1. Colorimetric Determination of Aluminum

With the exception of naphthenate additives, some greases, and a few special products, aluminum in petroleum products occurs in the range of 1 to 20 ppm, and is most frequently determined by a colorimetric procedure. The sample is prepared by the soft-ashing, wet-oxidation procedure as none of the aluminum compounds likely to be present is volatile. When the procedure given here was checked with mixes of aluminum stearate in lubricating oil, however, a loose powdery coating of aluminum oxide formed over the bottom and sides of the container to the height of the original oil surface. Recovery was incomplete because of mechanical losses of the light solid. When the samples were treated with about 5 drops of fuming sulfuric acid and heated on the steam plate for a few minutes before burning, the resulting ash was much more dense and crystalline, and was concentrated mainly at the bottom of the container.

Subsequent to the initial burning, the carbonaceous residue is oxidized in sulfuric acid with nitric acid and peroxide (see Chapter 2, Sections 2 and 3 for details). Aluminum sulfate separates from fuming sulfuric acid, but upon diluting with water and boiling it dissolves readily.

The procedure described here is essentially that of Strafford and Wyatt.[1] This procedure and others have been described by Sandell,[2] and the latter author also gives a comprehensive discussion of the aurintricarboxylate method. The procedure is useful for determining aluminum in crudes and residual fuel oils, and it is applicable to many other samples including used lubricating oils. As fuel oils contain a number of interfering elements—more, in fact, than any other sample likely to be handled—a discussion of the procedure in this application is in order.

The amount of sample taken should be sufficient to yield .025 to .15 mg of aluminum. Four samples of different grades of residual fuel oil were examined spectrographically, and the concentration ranges of the elements present and the number of milligrams of each in a 10-gram sample are tabulated in Table 3:1. These may be considered as typical results for this type of material.

TABLE 3:1

SPECTROGRAPHIC ANALYSES OF RESIDUAL
FUEL OILS

Element	Concentration, ppm	mg/10-gram Sample
Aluminum	3–7	0.03 –0.07
Calcium	7–12	0.07 –0.12
Copper	0.2–5	0.002–0.05
Iron	30–50	0.3 –0.5
Magnesium	7–11	0.07 –0.11
Nickel	50–65	0.50 –0.65
Silicon	5–20	0.05 –0.20
Sodium	20–50	0.2 –0.5
Vanadium	85–150	0.85 –1.5

It has been stated [1, 2] that calcium, magnesium, and phosphorus (as P_2O_5) do not interfere in amounts up to 10 mg, and that other common heavy metals, excluding iron and copper, do not interfere in amounts up to 1 mg. The effect of the amount of vanadium present here is questionable, but it can easily be removed with the iron and copper by extracting these three metals as cupferrates with chloroform. After these three elements are extracted it can be seen in Table 3:1 that the other metals are below the specified concentrations, and will not interfere. Chromium, if present, would interfere, but it will probably never be encountered.

If lead is present in sufficient concentration to interfere it will be precipitated as lead sulfate when the oxidized solution is diluted; and it can thus be filtered and discarded.

PREPARATION

1. REAGENTS:

Ammonium aurintricarboxylate: 0.2 gram in 100 ml of water. Let stand two or three days before using.

Gum arabic solution: 5 grams in 100 ml of water. Filter if necessary.

Buffer solution: Dissolve 156 g of ammonium acetate and 108 g of ammonium chloride in 1000 ml of water.

Standard aluminum solution: Dissolve 1.76 grams of $Al_2(SO_4)_3 \cdot K_2SO_4 \cdot 24H_2O$ in water, and dilute to 1000 ml in a volumetric flask (1 ml = 0.10 mg Al).

2. STANDARDIZATION:

1. Dilute the prepared standard aluminum solution tenfold to produce a solution containing 0.01 mg Al/ml. Transfer 2.5, 5.0, 7.5, 10.0, 12.5, and 15.0 ml of the dilute standard to 250-ml beakers, and include a blank.

2. From a graduated pipette add 4 ml of concentrated sulfuric acid to each beaker, and dilute to approximately 35 ml.

3. Cool the solutions, to each add one drop of methyl red indicator, and make it just alkaline with NH_4OH. Add 5N HCl until it is just acid and then add 5.0 ml excess, followed by two drops of bromine water to destroy the indicator. Finally, add 0.5 ml of 10% hydroxylamine hydrochloride to reduce the bromine, and dilute to volume in a 100-ml volumetric flask.

4. Transfer a 20-ml aliquot of each solution to 250-ml beakers, add 1.0 ml of 5% gum arabic solution, 5.0 ml of buffer solution, and 2.0 ml of ammonium aurintricarboxylate reagent. Mix, heat to boiling, and boil gently for 5 minutes. Cool to room temperature, transfer to 50-ml volumetric flasks, dilute to volume, and mix. Allow the solutions to

stand at least 5 minutes, and determine the transmittances at 525 mμ.

5. Plot the transmittances observed against the number of milligrams in the 50-ml volumetric flasks on semi-logarithmic graph paper. These amounts are one-fifth those originally taken.

PROCEDURE

If the sample is a heavy oil, heat it in a hot-water bath to reduce its viscosity, and mix by vigorous shaking.

1. Weigh a sample that will yield about 0.1 mg of aluminum, and transfer this to a dry 400-ml beaker. To the beaker add about 5 drops of fuming sulfuric acid, and heat the mixture on a steam plate for about 10 minutes, stirring thoroughly with a glass rod. Wipe the rod with a small piece of filter paper, and drop the paper into the beaker.

> NOTES: If the approximate content of aluminum is known the appropriate amount of sample can be estimated from the formula: $100/ppm = $ grams of sample.
> As mentioned in the discussion, the preliminary treatment with fuming sulfuric acid produces a denser ash than is obtained by direct ashing.

2. Place the beaker in a No. 2 tin can with the top cut out, support the can on a ring stand or tripod, ignite the sample by placing a luminous flame under the edge of the can, and allow the oil to burn without further heating. When the fire burns out remove the beaker from the can and burn the loose carbon from the walls of the beaker by directing a blast flame around the outside.

> NOTES: Oils with high ignition points may require continuous heating, but no more heat should be applied than is necessary to maintain burning. The bottom of the can should never be heated with a blast burner.
> The flame should not be directed at the bottom, nor across the top of the beaker, or loss of ash may result.

3. Cool the beaker, add about 15 ml of sulfuric acid, and fume strongly for about 10 minutes. Oxidize the carbonaceous material according to the principles covered in Chapter 2, Section 3, and reduce the final volume of sulfuric acid to about 4 ml. Cool, add about 25 ml of water, and boil for

about 5 minutes. If any insoluble material is present, filter the solution into a 125-ml separatory funnel, washing with two 5-ml portions of water to produce a final volume of about 35 ml.

> NOTES: The oxidation should be completed with perchloric acid to ensure that vanadium is quinquevalent for the subsequent cupferron-chloroform extraction.
> The solution must be boiled to ensure the complete solution of aluminum sulfate. The amount being handled here will not produce a visible precipitate, but the boiling should not be omitted. A cloudy solution after boiling indicates the presence of lead, silica, barium, or calcium.

4. Add 2.5 ml of a freshly-prepared 6% cupferron solution and mix. Extract with a 10-ml portion of chloroform, shaking for 30 seconds. Draw off and discard the chloroform layer, and extract with an additional 5-ml portion. Add 0.5 ml of cupferron solution, and extract successively with one 10-ml, and two 5-ml portions of chloroform, discarding all of the chloroform extracts.

> NOTE: If the determination of iron, copper, or vanadium is required, the chloroform extracts should be combined and reserved. Because of creeping, the chloroform cannot be evaporated directly without loss of the metallic cupferrates.

5. Drain the acid layer into a small beaker, and boil it until the residual chloroform is evaporated. Cool the solution, add one drop of methyl red indicator, and make it just alkaline with NH_4OH. Add $5N$ HCl until it is just acid and then add 5.0 ml excess, followed by two drops of bromine water to destroy the indicator. Finally, add 0.5 ml of 10% hydroxylamine hydrochloride to reduce the bromine, and dilute to volume in a 100-ml volumetric flask.

> NOTE: If any cupferron has been left in the solution the indicator color will not be destroyed.

6. Transfer a 20-ml aliquot to a 250-ml beaker, add 1.0 ml of 5% gum arabic solution, 5.0 ml of buffer solution, and 2.0 ml of ammonium aurintricarboxylate reagent. Mix, heat to boiling, and boil gently for 5 minutes. Cool to room temperature, transfer to a 50-ml volumetric flask, dilute to volume, and mix. Allow the solution to stand at least 5 minutes, and

determine the transmittance at 525 mμ. From the calibration prepared under *Standardization* determine the corresponding milligrams of aluminum.

NOTE: The amount of aluminum found in the 50-ml volumetric flask is one-fifth of the amount in the original sample.

2. Gravimetric Determination of Aluminum

The accurate determination of aluminum in most materials is difficult because it is not easily separated from a number of the other elements. It is indeed fortunate that the determination of aluminum in petroleum products is seldom required, and that in those few instances where it must be determined it is usually present alone. If it is possible to burn the sample without mechanical losses all of the aluminum initially present can be recovered in a direct ashing procedure. The insolubility of ignited aluminum oxide is, however, inconvenient if the ash must be dissolved for further treatment. In the analysis of aluminum soaps the material is sometimes ashed directly and weighed, after which the ash is treated with water and the undissolved material is reweighed—the latter being called aluminum oxide. Although this treatment may be suitable for products of this type it cannot be recommended for petroleum.

Aluminum compounds are used in the manufacture of grease, but the finished products are seldom analyzed for their content of aluminum. We shall consider the determination of aluminum in aluminum naphthenate, a commonly used additive in the petroleum industry. This product burns easily, but it is likely to spatter in the process and thus suffer mechanical losses. It is preferable, therefore, to oxidize the organic material in the wet way, and precipitate aluminum as the basic succinate according to the method of Willard and Tang.[3] As naphthenic acids are highly corrosive the presence of a small amount of iron is likely, and if the ignited aluminum oxide is colored a correction should be made for the ferric oxide present. The aluminum content may be in the range of one to three per cent, and one or two grams of sample will yield a satisfactory amount of alumina for a gravimetric determination.

It should be noted that if aluminum is to be determined in the presence of phosphorus the sample for the determination of aluminum should not be burned in a porcelain vessel because aluminum would thus be introduced into the solution to be analyzed. This would have to be considered only in the analysis of used lubricating oils, and if the colorimetric procedure given here is followed no such difficulty will be encountered.

PROCEDURE

1. Weigh a suitable sample and transfer it to a 400-ml beaker. If more than 2 grams is used follow with the procedure of soft-ashing, wet oxidation as described in Chapter 2, Section 2. With 2 grams or less, oxidize the sample in the wet way (Chapter 2, Section 3), finishing the oxidation with perchloric acid, and reduce the final volume of sulfuric acid to about 3 ml.

> NOTE: The amount of sample should be selected to yield a precipitate of less than 100 mg; 1.5–2 grams is satisfactory for aluminum naphthenate.

2. Cool the acid solution, dilute it to about 200 ml with water and boil it until the separated aluminum sulfate is in solution. Carefully neutralize the boiling solution with NH_4OH until a faint permanent turbidity or opalescence appears, and then add 5 grams of succinic acid. To the boiling solution add 5 grams of urea dissolved in a little water; place a small piece of filter paper under a stirring rod in the beaker to promote even boiling; and boil gently for 90 minutes. Water should be added from time to time during the boiling so that the final volume is about 150 ml.

> NOTES: Ten minutes of boiling is sufficient. If the solution is not clear after boiling, the presence of barium, calcium, lead, or silica is indicated. Cool the solution, set it aside for an hour, and filter it (Whatman No. 42). Wash with a dilute (1%) sulfuric acid solution.
>
> At least four grams of succinic acid must be present for complete precipitation; Willard and Tang recommend five grams. The addition of succinic acid should clear the solution; if it is still turbid add a drop or two of hydrochloric acid.

To ensure complete precipitation of aluminum this boiling period should not be shortened.

3. At the end of the boiling period remove the beaker from the heat, and filter the hot solution through Whatman No. 41 paper, washing the precipitate with a 1% solution of succinic acid made neutral to methyl red with NH_4OH.

4. If a layer of aluminum basic succinate adheres to the beaker, dissolve this in 2 ml of HCl, dilute to 50 ml, heat this solution to boiling, and neutralize it to methyl red with NH_4OH. Allow 2–3 minutes for the precipitate to flocculate, and filter through Whatman No. 41 paper, washing with the same solution as in step 3. Discard both the filtrates and the washings.

Combine the two filter papers containing the precipitates in a tared crucible, dry them, char, and ignite for one hour at 1100°C. Cool and reweigh, recording the weight of Al_2O_3.

NOTES: As the amount of aluminum is small it is precipitated with ammonia.

If the precipitate is significantly discolored by iron it should be fused with potassium pyrosulfate and the iron determined colorimetrically; see Chapter 12, Section 4.

References

1. N. STRAFFORD and P. F. WYATT, "The Determination of Small Amounts of Aluminum by the Aurintricarboxylate Method," *Analyst*, 72:54 (1947).
2. E. B. SANDELL, *Colorimetric Determination of Traces of Metals*, 2nd Edition, Interscience, New York (1950).
3. H. H. WILLARD and N. K. TANG, "Quantitative Determination of Aluminum by Precipitation with Urea," *Ind. Eng. Chem., Anal. Ed.*, 9:357 (1937).

Chapter 4

ARSENIC

Very little information has been published concerning the amount and nature of arsenic in crude oils, although phosphides and sulfides have been isolated. In those samples that have been examined, the arsenic content is less than one part per million. In common with other metals that form reducible oxides,[1] arsenic is poisonous to cracking catalysts, and it is particularly damaging to the platinum catalyst used in re-forming processes, as it apparently poisons platinum mol for mol. It is in the charge stocks for this process that arsenic is most frequently determined.

In plants provided with units for reducing the sulfur, nitrogen, lead, and arsenic in the feed, the amount of arsenic is of little importance. Molybdenum hydroforming catalysts [2] are effective for removing the four elements named above, and as much as 1000 ppb of arsenic can be tolerated for relatively long periods. On the other hand, if the naphtha to be re-formed is fed directly to the platinum catalyst the content of arsenic must be below 10 ppb for economical operation. This means that straight-run naphtha must be used because its arsenic content is usually small. In cracked naphtha the concentration of arsenic may be 200 ppb or higher, depending upon the source of the crude and other factors, and these naphthas must be pretreated before being used for re-forming charge stock.

Table 4:1 shows the sensitivity of the arsenic content to the material of containers and time of storage. The tabulated determinations were made by the procedure that follows, after extraction of arsenic with hypochlorite and sulfuric acid, and wet oxidation of the extracts. Samples I and II are cracked naphthas, and sample III is straight-run naphtha.

TABLE 4:1

EFFECT OF CONTAINER ON ARSENIC CONTENT

Sample	Container	ppb Arsenic after storage			
		1 hour	48 hours	7 days	14 days
I	Can	22	46	32	33
	Soft Glass	21	57	37	38
	Pyrex	29	30	32	—
II	Soft Glass	59	—	52	—
III	Soft Glass	Nil	—	2.8	—

The most probable arsenic content of sample I appears to be about 30 ppb. As can be seen in Table 4:1 there is an initial rapid reduction of the arsenic concentration in both the can and the soft glass bottle. Upon standing, the concentration increases markedly, and then falls off progressively, stabilizing after about a week. This behavior may be correlated with the content of basic nitrogen; see Chapter 17 for determination of basic nitrogen.

The basic nitrogen contents of the samples are as follows: Sample I, 130 ppm; Sample II, 202 ppm; Sample III, less than 0.1 ppm. Arsenic trioxide, an amphoteric component of glass, is more acidic than basic; consequently it may participate in an acid-base reaction with nitrogen bases in the naphtha, and thus increase the arsenic content as time passes. Upon standing, or under the influence of light, the nitrogen bases polymerize. This change may again release arsenic, which would return to the walls of the container and thus decrease the arsenic content of the sample.

Samples with basic nitrogen contents in the range of 100–200 ppm are pale yellow when freshly drawn. As polymerization proceeds, the color darkens progressively over a period of two weeks, and the naphtha eventually becomes almost black. The extracts also become more difficult to oxidize, and more likely to foam the longer the sample has aged.

It should be mentioned that as mixes of triphenylarsine in naphtha lose arsenic to the walls of the container, the arsenic content becomes gradually less. If these mixes are stored in ordinary glass bottles, and both the naphtha and the bottle are extracted, considerably more arsenic is recovered than was put into the bottle. The practice of extracting sample bottles, which is sometimes recommended, produces results that are much too high. Because as much as 40 micrograms of arsenic may be removed from the walls of a clean soft glass bottle by the extraction procedure described here, this practice should not be followed. Evidently the only satisfactory container for samples is a Pyrex bottle.

The precise determination of small amounts of arsenic is extremely difficult. The procedure that follows, however, was applied to a sample in which arsenic had been determined by several different laboratories and by a number of methods, and the content was established at 34–36 ppb. The procedure to be described gave 34 and 38 ppb as duplicate results. If equipment is available the analyst may prefer the method described by Maranowski et al.[3] for some applications. Their method employs a wet-oxidation procedure to prepare the sample, and a reflectometric measurement of the stains produced in the Gutzeit method of evolution.[4]

1. Colorimetric Determination of Arsenic

As the concentrations of arsenic encountered in petroleum are always very small, a very sensitive method for determining it is necessary. The original Gutzeit procedure has been modified in many ways, one of the major improvements being the application of developing the color of heteropoly molybdenum blue for the final determination. Comprehensive summaries of the Gutzeit method have been given by How,[5] and by Jacobs and Nagler.[6] The principal difficulty with the original method is in establishing uniform conditions for producing the stained strips. By oxidizing the evolved arsine to arsenate, and measuring the blue color produced with molybdate and a reducing agent, most of the problems of the arsine evolution process are avoided; the rate of evolution need not be closely controlled, and the arsine is quantitatively recovered in one absorber.

There are virtually no problems of interference to be considered. Phosphate as such does not interfere as it is not reduced to phosphine by zinc. If phosphides were present, phosphine would be evolved and cause high results, but the oxidation step precludes this possibility. In the amounts normally encountered, chromium, iron, molybdenum, and lead are without effect, except that the first three increase the rate of evolution of hydrogen. Easily reduced metals, such as cobalt, copper, and nickel, produce low results if several milligrams are present. With the exception of lead and nickel (seldom present in excess of a fraction of a part per million), none of these elements is found in re-forming charge stocks.

There is a difference of opinion as to the effect of charring during the wet oxidation. Maranowski *et al.*,[3] and Jacobs and Nagler [6] report that arsenic is lost if charring occurs, whereas How,[5] and Barnes and Murray [7] state that the final result is not affected. As usual in such disagreements, the results probably depend upon the type of material being investigated; with petroleum products, moderate charring is not significant.

The method given here is a composite of several procedures: The arsenic standard is prepared as described by the Association of Official Agricultural Chemists,[8] arsenic is extracted, and arsine evolved, according to a procedure of Universal Oil Products,[9] and the molybdate reagent and acidity for color development are as given by Jacobs and Nagler.[6] The iodine-bicarbonate absorbing solution is similar to that recommended by Rogers and Heron,[10] but less bicarbonate is used.

PREPARATION

1. REAGENTS:

Ammonium Molybdate Solution: Dissolve 25 grams of ammonium molybdate in 300 ml of water. Dilute 150 ml of 1:1 H_2SO_4 to 200 ml, and add this to the ammonium molybdate solution.

Ammonium oxalate: Saturated solution.

Clorox: Commercial household bleach.

Potassium iodide: Dissolve 15 grams of KI in 100 ml of water.

Sodium bicarbonate-iodine: Dissolve 4.2 grams of $NaHCO_3$ in 100 ml of water, and add 200 ml of $N/10$ iodine solution.

Sodium metabisulfite: 5% solution freshly prepared.

Stannous chloride: Dissolve 20 grams of $SnCl_2 \cdot 2H_2O$ in 50 ml of HCl.

Stannous chloride, dilute: Dilute 1.0 ml of the above solution to 200 ml with water. Prepare fresh each day.

Sulfuric acid, 70%: Add 3 volumes of concentrated H_2SO_4 to 2 volumes of water.

Sulfuric acid, 2N: Dilute 55 ml of concentrated H_2SO_4 to one liter.

Standard arsenic solution: Dissolve 0.132 gram of As_2O_3 in 25 ml of 20% NaOH solution. Saturate with CO_2, and dilute to one liter with water. 1 ml = 0.10 mg As. Prepare dilute standards immediately before use, as required.

2. STANDARDIZATION:

1. Dilute 10.0 ml of standard arsenic solution to 1000 ml; transfer 1.0, 3.0, 5.0, 7.0, and 10.0 ml of dilute standard to 25-ml volumetric flasks, and include a blank.

NOTE: The milliliters of dilute standard are equivalent to micrograms of arsenic; i.e., 1.0, 3.0, 5.0, 7.0, and 10.0.

2. To each flask add 3.0 ml of sodium bicarbonate-iodine solution, dilute to about 15 ml and let stand 5 minutes. To each flask add 5.0 ml of 2N H_2SO_4 and 1.0 ml of ammonium molybdate solution, mixing between additions. Add 5% sodium metabisulfite drop by drop until the iodine is reduced, and then add 1.0 ml of the dilute stannous chloride solution.

3. Dilute to volume, mix, let stand 5 minutes and determine the transmittance at 700 mμ using 50-mm cuvettes. Plot the percentage of transmittance observed against the micrograms of arsenic taken on semi-logarithmic graph paper.

4. Dilute 10.0 ml of the standard arsenic solution to 100 ml; transfer 1.0, 2.0, 3.0, 4.0, and 5.0 ml of the dilute standard (10, 20, 30, 40, and 50 micrograms) to 25-ml volumetric flasks, and proceed as described except that the transmittances are determined in 13-mm cuvettes.

Two calibrations are thus obtained, the first covering the range from 0 to 10 micrograms and the second from 0 to 50

micrograms. Either curve may be used, depending on the
depth of color obtained from the sample.

PROCEDURE

1. Transfer a suitable sample of naphtha to a separatory
funnel of appropriate size, add 15 ml of Clorox and shake
it vigorously for 5 minutes. Allow the layers to separate,
and drain the Clorox layer into a 300-ml Kjeldahl flask. Add
10 ml of 70% H_2SO_4 to the funnel, shake it for 5 minutes,
and combine the acid layer with the Clorox extract. Repeat
this, using 5 ml of H_2SO_4. Finally extract with a 10-ml portion of water. The stem of the funnel should be rinsed with
water after each extraction, and all rinsings and extracts
should be combined in the Kjeldahl flask. A blank must be
carried through this entire procedure.

> NOTE: Cracked naphthas may contain from 25 to 250 ppb As;
> straight run naphthas contain less, usually 1 to 10 ppb. Treated
> naphtha will probably contain less than 1 ppb. The amount of the
> sample should be selected to yield 2–10 micrograms of arsenic.
> (If a sample contains 10 ppb, 500 g will yield 5 micrograms, etc.).

2. Transfer the flasks to a hood, add 25 ml of HNO_3 to
each flask, mix by swirling, and allow them to stand at least
30 minutes. Add a few carborundum crystals (or Alundum
chips) to prevent bumping, place the flasks on electric heat-
ers equipped with variable controls, and warm them gently
to start the oxidation. As the initial reaction slows, increase the heat, being careful not to raise the temperature
so rapidly that the mixture foams excessively.

> NOTE: If multiple determinations are to be made, it is convenient
> to set up several heaters side by side.

3. When the initial vigorous reaction subsides and the
mixture begins to darken, add nitric acid in small increments until the solution becomes straw-colored. Increase the
heat to expel excess of nitric acid and when the temperature
of the solution reaches the point of incipient fuming of the
sulfuric acid, allow about 5 ml of 30% H_2O_2 to flow slowly
down the neck of the flask. If the solution is still colored,

repeat the peroxide treatment until there is no further evolution of nitrogen oxides.

> NOTES: It is advisable not to allow the digestion mixture to get darker than a deep brown during the oxidation lest arsenic be lost under these reducing conditions.
> If the solution darkens after nitric acid is expelled, add more nitric acid.
> It is best to use a grade of hydrogen peroxide that contains no preservative.

4. Raise the temperature until heavy fumes of sulfuric acid appear, allow the solution to fume for about 5 minutes, and then cool it. When the solution is fairly cool add very slowly 15 ml of saturated ammonium oxalate solution. Heat to boiling, fume for a period of 5 minutes, and cool to room temperature.

> NOTE: Ammonium oxalate is added to remove residual NO_2, although this may not be neecssary. The solution must be fairly cool to avoid violent reaction.

5. Add to the flask about 10 ml of water, and transfer the solution to the arsine-generating apparatus (Figure 1),

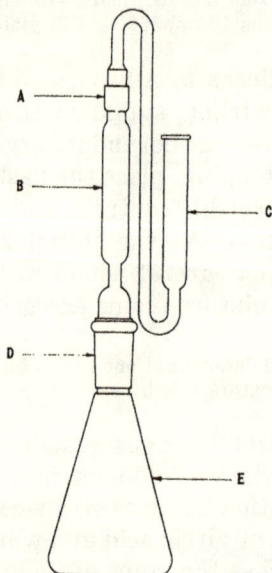

FIG. 1. ARSINE GENERATOR AND ABSORBER ASSEMBLY
a) Standard taper joint 7/25; b) Guard tube; c) Absorber with glass beads; d) Standard taper joint 19/38;
e) 125-ml Erlenmeyer flask.

diluting it in the process to about 75 ml. Cool to room temperature, add 5 ml of potassium iodide solution, swirl and allow to stand for 15 minutes, then add 1 drop of stannous chloride solution.

NOTES: Mark the volume on the flask with a wax pencil.
Potassium iodide reduces quinquevalent arsenic to the trivalent state increasing the rate of evolution of arsine.
The presence of stannous ion is essential for the evolution of arsine; very little is evolved in its absence.

6. Fill the guard tube about two-thirds full with glass wool, wash with a few portions of water and then with 20% NaOH, washing off any excess reagent from the outside of the tube and the inside of the ground glass joint. Put glass beads into the special absorber, attach it to the guard tube, and add 3.0 ml of sodium bicarbonate-iodine solution.

NOTE: Sodium hydroxide absorbs any H_2S or SO_2 formed in the reaction. The preparation of the tube is aided by applying suction to the bottom during the rinsing.

7. Add to the Erlenmeyer flask about 5 grams of 20-mesh granular zinc, and immediately attach the guard tube and absorber assembly. Allow the evolution to proceed for $1\frac{1}{2}$ hours, then transfer the contents of the absorber to a 25-ml volumetric flask, and wash the absorber with several 2-ml portions of water. Add 5.0 ml of 2N H_2SO_4, and 1.0 ml of ammonium molybdate reagent, mixing between additions. Reduce the iodine by adding drops of a 5% solution of sodium meta-bisulfite; then add 1.0 ml of dilute stannous chloride, dilute to volume, and mix.

NOTES: Before use, the zinc should be pickled briefly with hydrochloric acid, washed, and stored under water. It is conveniently measured with a calibrated scoop or spoon.
A small funnel with a plug of glass wool may be used to separate the beads, and facilitate the transfer of the solution and washings.

8. Allow the mixture to stand 5 minutes, and determine the transmittance at 700 mμ. From the appropriate calibration curve read the corresponding micrograms of arsenic, correct for the reagent blank, and calculate the concentration of arsenic in the original sample.

NOTE: Reagent blanks normally amount to about 2 micrograms, most of which comes from the zinc.

References

1. G. A. MILLS, "Aging of Cracking Catalysts", *Ind. Eng. Chem.*, 42:182 (1950).
2. B. H. DANZIGER and J. R. MILLIKEN, "Molybdenum in Petroleum Refining", *Refinery Engineer*, Nov., Dec., 1956.
3. N. C. MARANOWSKI, R. E. SNYDER, and R. O. CLARK, "Determination of Trace Amounts of Arsenic in Petroleum Distillates", *Anal. Chem.*, 29:353 (1957).
4. M. GUTZEIT, *Pharm. Ztg.*, 24:263 (1879).
5. A. E. HOW, "Microdetermination of Arsenic", *Ind. Eng. Chem., Anal. Ed.*, 10:226 (1938).
6. M. B. JACOBS and J. NAGLER, "Colorimetric Microdetermination of Arsenic". *Ind. Eng. Chem., Anal. Ed.*, 14:442 (1942).
7. J. W. BARNES and C. W. MURRAY, "Accuracy of the Gutzeit Method for the Determination of Minute Quantities of Arsenic", *Ind. Eng. Chem., Anal. Ed.*, 2:29 (1930).
8. Official Methods of Analysis of the Association of Official Agricultural Chemists, 7th Ed. (1950).
9. Universal Oil Products Method Number H–296–53.
10. D. ROGERS and A. E. HERON, "The Determination of Small Amounts of Arsenic", *Analyst*, 71:414 (1946).

Chapter 5

BARIUM

Barium occurs naturally in very few crude oils but it is occasionally introduced as a contaminant when drilling muds composed of bentonite clays weighted with barite ($BaSO_4$) are used in removing the oil from the ground. Barium introduced in this manner is insoluble and nonvolatile, and as the amount would in any case be small, it is not ordinarily significant in refining operations and determination of it is rarely required. The element has not been reported as an oil-soluble component of crude oil.[1]

Although barium is a common additive in lubricating oil, there is one application where its presence is undesirable. It is one of several metals whose action is deleterious to the silver bearings and the wrist pin bushings in diesel locomotives. During the manufacture of diesel engine lubricating oil every effort is made to avoid contamination, and special mixers are usually reserved for compounding it. As the concentration of barium is less than one part per million in an acceptable product, it must be determined by spectrographic methods. Special sampling is required for the determination in locomotive crankcase oil of metals introduced by wear[2] to ensure accurate results for evaluating engine wear. Calkins and White[3] have described a spectrographic determination of barium in additives and lubricating oils. Because barium is almost invariably determined as a component of lubricating oils, the following discussion will be confined to this application.

Common forms of barium additives are sulfonates and phenates, as well as various undisclosed formulations. Brauns et al.[4] have described an ion-exchange procedure for determining barium and sulfur in organic barium sulfonates which is very convenient for handling those additives that are water-soluble. Elements commonly associated with barium in newly compounded lubricating oils are calcium, phosphorus, sulfur, and zinc, and the gravimetric determination of barium in these oils is not difficult. With used lubricating oils, however, the analysis is often complicated by the presence of additional elements such as aluminum, chromium, copper, iron, lead, nickel, and silicon.

Of these contaminants in used oils, aluminum, chromium, copper, iron, and nickel are introduced by wear from bearings, piston rings, etc.; and as their concentration is usually small high concentrations may indicate corrosion. The presence of lead may be an indication of worn piston rings and it is often accompanied by "blow-by" and dilution by fuel. Gasolines with high sulfur content produce engine deposits containing as a major component $PbSO_4 \cdot PbO$, some of which may reach the crankcase. Silica in lubricating oil usually enters as dust through failure of air filters. Chromium in diesel locomotive oil may indicate defective gaskets, which allow chromate-treated cooling water to enter the lubricating system. Locomotive oils, however, never contain enough barium to permit gravimetric determination.

In devising a procedure for the gravimetric determination of barium, the coprecipitation of various ions with barium sulfate must be considered. The results of several investigations of this phenomenon [5, 6, 7] may be summarized as follows:

Chromium (III) and iron (III) are coprecipitated as complex ions rather than by absorption;

Calcium tends to be carried down because of the moderate solubility of calcium sulfate (Paneth-Fajans-Hahn absorption rule);

Barium chromate and barium sulfate are isomorphous, as are barium sulfate and lead sulfate, and mixed crystals are formed;

Zinc is not coprecipitated with barium sulfate but it is isomorphous with lead sulfate, and in the presence of the latter zinc contaminates the mixed precipitate, as does copper;

Nickel and phosphate are ordinarily insignificant.

As the analysis of new oils is much simpler than the analysis of used oils, two separate procedures are given here. The procedure for new lubricating oils takes account of the possible presence of calcium, phosphorus, sulfur, and zinc. The procedure for used oils provides for the determination of barium in the presence of these elements as well as aluminum, chromium, copper, iron, lead, nickel, and silicon.

1. Determination of Barium in New Lubricating Oils

It was remarked in Chapter II that wet oxidation is not a convenient method of preparing a sample for the determination of barium or calcium. The reason is that these two metals are frequently present together, and upon prolonged fuming with sulfuric acid, calcium sulfate separates in an anhydrous form which cannot be redissolved unless metathesized with sodium carbonate. Calcium sulfate that is ignited in a furnace does not behave this way, but it dissolves readily in dilute acid even after prolonged ignition. Thus, if a sample containing barium and calcium is wet-oxidized, and the insoluble material is treated as barium sulfate, the result for barium is always too high because of the coprecipitated calcium sulfate. If the wet oxidation is prolonged, and the volume of sulfuric acid decreases, the amount of calcium sulfate that is rendered insoluble increases, and as the solution approaches dryness practically all the calcium comes out of solution and cannot be redissolved.

To illustrate the magnitude of the error, two synthetic concentrates were prepared by dissolving known amounts of barium phenate, calcium sulfonate, and an alkyl zinc dithiophosphate in a diluent. The results obtained by soft ashing and wet oxidation, and by direct ashing are shown in Table 5:1. The calcium was determined in the filtrates by the oxalate-permanganate volumetric procedure with double precipitations. Although these synthetic additives do not correspond to any compounding concentrate in actual use, the results indicate the importance of selecting an appropriate method for destroying the associated organic material.

TABLE 5:1

COPRECIPITATION OF CALCIUM WITH BARIUM SULFATE

Sample	% Present	% Found	
		Soft Ash, Wet Oxidation	Direct Ash
I			
Barium	4.36	4.50, 4.52, 4.55, 4.56	4.36, 4.37
Calcium	0.93	0.83, 0.79, 0.77, 0.70	0.92, 0.93
Phosphorous	0.9	......	......
Sulfur	2.7	......	......
Zinc	0.9	......	......
II			
Barium	1.75	1.77, 1.77	1.72, 1.73
Calcium	1.13	1.06, 1.08	1.12, 1.13
Phosphorous	0.5	......	......
Sulfur	1.5	......	......
Zinc	0.5	......	......

From the tabulated results, it is apparent that the best procedure to use for preparing the sample is direct ashing, and it is also seen that moderate amounts of phosphorus, sulfur, and zinc are without effect in the determination of both barium and calcium. It should also be remarked that after direct ashing, the ashes of both of these samples dissolved completely in dilute hydrochloric acid, showing that no barium sulfate was formed in the incineration and ignition. Some barium additives, however, containing sulfurized sperm oil, form barium sulfate upon ignition, although in the presence of carbon this is partially reduced to barium sulfide. The latter dissolves in dilute hydrochloric acid and the barium sulfate remains as an insoluble residue. In any event the ash is readily loosened from the dish, and any calcium sulfate or sulfide that may have been formed dissolves.

To precipitate the barium remaining in solution, a procedure described by Wagner and Wuellner,[8] which involves the hydrolysis of sulfamic acid, has been modified and applied. This method produces a homogeneous precipitate of barium sulfate, practically free from calcium, iron, and phosphate, and it is especially appropriate for handling lubricating oils in which these elements are frequently encountered. Zinc does not coprecipitate in any case, so it presents no problem.

PROCEDURE

1. Weigh an appropriate sample and transfer to a Coors 2/0A porcelain dish. Heat the dish with a soft flame until the oil ignites, and allow it to burn freely until the fire is extinguished. Burn off the loose carbon by careful application of the flame, and finally ignite the dish in a furnace at about 800°C until most of the carbon is oxidized.

NOTES: A convenient amount of $BaSO_4$ to handle is about 100 mg; 12 grams of a sample containing 0.5% barium would produce this amount.

It is best to use porcelain ware for compounded oils containing phosphorous, sulfur, or zinc.

It is unnecessary to take time to burn all the carbon in this first ignition as it will be burned completely when the $BaSO_4$ is finally ignited.

2. Remove the dish from the furnace and cool it to room temperature. Moisten the ash with about 10 ml of water, add 2 ml of HCl, and warm it on the steam plate until the ash has loosened or dissolved. Transfer the solution and any undissolved residue to a 400-ml beaker, policing the dish if necessary. Dilute to about 250 ml, add 1 gram of NH_2SO_3H, heat to boiling, and boil for 30 minutes.

NOTES: The residue is moistened with water first to prevent spattering when the acid is added.

The solution is boiled to ensure complete precipitation; no digestion is required.

3. Cool the solution to room temperature and filter through Whatman No. 40 paper, washing with water. Transfer the paper and precipitate to a tared crucible, smoke off the paper in a radiator, and ignite for 45 minutes at 900°C. Cool the crucible in a desiccator and reweigh it, calculating the percentage of barium in the original sample.

NOTES: Ignition temperatures lower than 900°C give high results.[8]

The gravimetric factor for barium from barium sulfate is 0.5885.

2. Determination of Barium in Used Lubricating Oils

The presence in used oils of easily reducible metals which attack platinum, and the fact that several metals are rendered insoluble by ignition, and thus require carbonate fusion to redissolve them, precludes the direct ashing method. Therefore, in spite of the objections to the method of soft ashing and wet oxidation mentioned in the preceding section, it must be used in this instance. The amount of sample should be selected so that the amount of insoluble material in the sulfuric acid solution at the conclusion of the wet oxidation will not exceed 100 mg. The appropriate amount may be estimated from a preliminary rough determination of total ash, taking account of the fact that some of the ash probably will be soluble. The weighed sample is transferred to a beaker, burned, and wet-oxidized in the usual way (Chapter 2, Section 3), the oxidation being completed with perchloric acid.

It is unlikely that any single sample will contain all twelve of the elements that the following procedure takes into account, and it would be tedious to consider in detail the various possible combinations. In the procedure described here, however, the general distribution of the elements between residues and filtrates will be outlined.

At the conclusion of the wet oxidation, before the sulfuric acid is diluted, the following conditions prevail: Copper, phosphate, and zinc (in the absence of lead), will be essentially all in solution; aluminum, barium, iron, lead, and silica will be practically all in the residue; the residue will be more or less contaminated by calcium, chromium, and nickel, and in the presence of lead, by zinc and copper. Aluminum, iron, and nickel form insoluble sulfates in fuming sulfuric acid which redissolve when diluted and boiled; the resolution of iron is facilitated by reduction with hydrazine or hydroxylamine, and this also decreases its coprecipitation with barium sulfate. In fuming sulfuric acid, chromium separates as $Cr_4H_2(SO_4)_7$ which does not redissolve. This compound is slowly oxidized to chromate by boiling perchloric acid, and is coprecipitated to some extent as barium chromate although the reducing agent minimizes this effect.

Upon diluting, reducing, boiling, and filtering, practically all the aluminum and iron dissolve and pass into the filtrate. The residue contains essentially all the barium, lead, and silica, significant amounts of calcium and chromium, and rather small amounts of copper and zinc (the latter two only if lead is present).

The washed residue is next leached with hot ammonium acetate solution to dissolve lead sulfate, and concomitantly the small amounts of copper and zinc coprecipitated with the lead sulfate. The separation by ammonium acetate has been investigated by Scott and Alldredge,[9] who found it satisfactory for barium and lead but not very effective for calcium and lead. It is therefore important that the volume of sulfuric acid should not be allowed to decrease too much during the wet oxidation in order to minimize contamination of the residue by calcium. To ensure that this leaching process will be effective, it is essential that the weight of the residue should not exceed 100 mg.

The leached residue is next fused with a mixture of sodium carbonate and potassium nitrate. This oxidizes chromium to chromate, converts silicic acid to soluble silicate, and metathesizes barium and calcium sulfates to carbonates. The cooled flux is taken up in water, and as the mixed carbonates are filtered, chromate, silicate and sulfate pass into the filtrate. Finally, the mixed carbonates are dissolved in dilute hydrochloric acid, and the barium is precipitated as in Section 1 (New Oils) by the hydrolysis of sulfamic acid.

PROCEDURE

1. Weigh a suitable sample and transfer it to a dry 400-ml beaker. Reduce the sample to a soft ash and wet-oxidize the residue as described in Chapter 2, Sections 2 and 3, finishing the oxidation with perchloric acid to a final volume of about 5 ml.

NOTES: The amount of sample should be enough to yield not more than 100 mg of material insoluble in sulfuric acid.

If the wet oxidation is unduly prolonged, an excessive amount of calcium sulfate contaminates the residue.

2. Dilute the cool acid solution to about 150 ml, add one gram of hydrazine (or hydroxylamine) sulfate and boil gently for 10 minutes. Digest the solution for 1 hour on the steam plate, cool it to room temperature, and filter through Whatman No. 42 paper. Wash the residue with water and discard the filtrate and washings.

NOTE: Increasing the time of digestion decreases coprecipitation but after 3 hours there is little effect; one hour is the minimum time required.

3. Pour 50 ml of hot 50% NH_4Ac through the paper, making sure that all the residue is thoroughly saturated, and then wash thoroughly with water. Transfer the paper and precipitate to a platinum crucible and place in a radiator, heating until all of the carbon has been burned.

NOTES: The ammonium acetate solution should be poured over the residue and paper in 10-ml portions, and each portion should be allowed to drain away completely before the next is added.
This residue should not be ignited and, as ignition renders chromium less soluble, it is undesirable to do so.

4. Using a small glass rod, mix the residue with 4–5 grams of a fusion mixture composed of 10 parts anhydrous Na_2CO_3 and 1 part KNO_3, cover the crucible and fuse for 15 minutes at the full heat of a blast burner.

NOTE: If a large amount of chromium is present, as evidenced by the color of the residue, the amount of flux should be increased, and the heating continued for 30 minutes.

5. Cool, dislodge the solidified mass from the crucible by bending it slightly, and transfer it to a 400-ml beaker, dissolving in water any particles of flux that remain in the crucible. Add about 100 ml of water to the beaker, and stir until the mass has disintegrated and the excess flux has dissolved. Filter the residue on Whatman No. 42 filter paper, washing with a few portions of a saturated solution of $BaCO_3$, and finally with one portion of water.

NOTE: Significant amounts of barium carbonate dissolve if washed with water.

6. Place the original beaker under the funnel, punch a hole in the paper with a small glass rod, and wash as much

as possible of the residue into the beaker with water. To 15 ml of water add 2 ml of HCl, heat this to boiling and pour it over the paper, receiving the solution in the same beaker. Finally, wash the paper thoroughly with water, fold it into a compact packet, and reserve.

NOTES: Hydrochloric acid must be washed out completely as filter paper wet with this acid burns with difficulty.

The metathesis of barium sulfate is not quite complete in one fusion, and as the unchanged barium sulfate is finely divided Whatman No. 42 paper must be used.

7. Dilute the contents of the beaker to 250 ml, heat to boiling, slowly sift 1 gram of NH_2SO_3H into the solution, and boil it for 30 minutes.

NOTE: As there is no more than a slight haze present after a successful fusion, this solution should be practically clear.

8. Cool the solution and filter it through Whatman No. 40 paper, washing with water. Transfer the paper and precipitate to a tared platinum crucible and add the filter paper reserved in step 6. Cover the crucible, smoke off the paper in a radiator, and ignite the residue for 45 minutes at 900°C. Cool the crucible and reweigh it, calculating the amount of barium in the original sample.

References

1. M. C. K. JONES and R. L. HARDY, "Petroleum Ash Components and Their Effect on Refractories," *Ind. Eng. Chem.*, 44:2615 (1952).
2. "Sampling Diesel Locomotive Lubricating Oil for Spectrographic Analysis," *Bulletin Am. Soc. Testing Materials*, No. 208:24 (1955).
3. L. E. CALKINS and M. M. WHITE, "Analyze Additive Lubricants in Minutes Instead of Hours with Spectrographic Method," *National Petroleum News*, 38, No. 27:519 (1946).
4. F. E. BRAUNS, J. B. HLAVA, and H. SEILER, "Determination of Sulfur and Barium in Organic Barium Sulfonates," *Anal. Chem.*, 26:607 (1954).
5. L. WALDBAUER and E. GANTZ, "Quantitative Spectrographic Studies of Coprecipitation," *Ind. Eng. Chem., Anal. Ed.*, 5:311 (1933).
6. L. WALDBAUER, F. W. ROLF, and H. A. FREDIANI, "Spectrographic Studies of Coprecipitation," *ibid.*, 13:888 (1941).
7. W. B. MELDRUM, W. E. CADBURY, JR., and C. E. BRICKER, "Coprecipitation of Chromate with Barium Sulfate," *ibid.*, 15:560 (1943).

8. W. F. WAGNER and J. A. WUELLNER, "Homogeneous Precipitation of Barium Sulfate by Hydrolysis of Sulfamic Acid," *Anal. Chem.*, 24:1031 (1952).
9. W. W. SCOTT and S. M. ALLDREDGE, "Investigation of the Ammonium Acetate Separation of Sulfates of Lead, Barium, and Calcium," *Ind. Eng. Chem., Anal. Ed.*, 3:32 (1931).

Chapter 6

BORON

Boron has been reported in only a few crude oils and in these, only in trace amounts.[1] It is a relatively unimportant element in the petroleum industry, practically the only applications being the use of borax in a very limited number of specialty products, and the addition of glycol borates to gasoline to control surface ignition. These latter compounds are less effective, however, than other additives, such as tricresyl phosphate, and are therefore not extensively used. If they are used with tetraethyllead fluids that contain ethylene chloride as a lead scavenger, they have the further disadvantage of causing the burning of valves.

As a gasoline additive, boron is generally present in concentrations of 10 to 50 ppm, and it is extractable with water. If a flame spectrophotometer is available it offers the most convenient method of determining this element. Dean and Thompson[2] have studied the method and their article should be read for pertinent details. A colorimetric procedure given here utilizes quinalizarin; and for the determination of larger concentrations an alkalimetric procedure is provided. There is seldom any occasion to apply either method, but both are included for the sake of completeness.

1. Colorimetric Determination of Boron

The concentration range of boron when used to inhibit surface ignition in gasoline is in the range of 10 to 50 ppm, and by using suitable dilutions, it can be determined by the very sensitive quinalizarin method.[3] The element is added as a water-soluble glycol borate which hydrolyzes readily to boric acid. As will be seen, the dilution factor is rather large, and the precision of the determination is affected accordingly. Also, the necessity of using concentrated sulfuric acid as a working medium is inconvenient.

PREPARATION

1. REAGENTS:

> *Quinalizarin:* Dissolve 10 mg in 250 ml of concentrated sulfuric acid. Allow this to stand for one day with occasional shaking.
>
> *Standard Boron Solution:* Dissolve 0.572 grams of boric acid in water and dilute to one liter. 1 ml = 0.10 mg B.

2. STANDARDIZATION:

1. Transfer 2.0, 4.0, 6.0, and 8.0 ml portions of the standard boron solution to 100-ml volumetric flasks and dilute to volume with water. Transfer a 1.0 ml aliquot of each dilution to a dry 50-ml beaker and add to each beaker 10.0 ml of quinalizarin reagent, swirling during the addition to dissipate the heat. Prepare a blank solution containing 1.0 ml of water and 10.0 ml of reagent to be used as reference.

> NOTES: The 1-ml aliquots contain respectively 2, 4, 6, and 8 micrograms of boron.
>
> The color of the blank solution resembles that of permanganate, and in the presence of boric acid the reagent turns blue. As the blue is measured in the presence of an excess of color from the reagent, the spectrophotometer is set at 100% transmittance, using the blank solution as reference.

2. Cool the solutions in a water bath and allow them to stand for 20 minutes to develop the full color. Transfer the blank solution to a clean, dry 13-mm cuvette and set the transmittance at 100% at a wavelength of 620 mμ. Determine the transmittances of the remaining solutions, cleaning and drying the cuvette after each measurement, and plot the transmittances against the corresponding micrograms of boron on semi-logarithmic cross-section paper.

> NOTE: The color is stable for at least 24 hours after full development.

PROCEDURE

1. Transfer a 10-ml portion of gasoline to a 60-ml separatory funnel and extract with two 20-ml portions of water, shaking for two minutes in each extraction, and then transfer the aqueous layers to a 100-ml volumetric flask. Dilute to volume, mix, and transfer a 1.0-ml aliquot to a clean, dry 50-ml beaker. Add 10.0 ml of quinalizarin reagent while swirling the funnel and cool it to room temperature in a water bath.

2. Allow the extraction to stand at least 20 minutes, then determine the transmittance at 620 mμ, using a blank solution prepared as described in the preceding section. From the prepared calibration curve, read the corresponding micrograms of boron and multiply this number by 100 to obtain the total amount extracted from the sample.

3. Calculate the parts per million of boron in the sample from the volume taken and the specific gravity of the gasoline.

2. Alkalimetric Determination of Boron

When determining larger amounts of boron (4–6%) in gasoline additives or other organic compounds, a very convenient procedure is that given by Burke.[4] Organic material is destroyed by ignition in an oxygen bomb [5] containing a solution of sodium carbonate. The resulting alkaline solution is transferred to an Erlenmeyer flask, acidified, and the resultant carbon dioxide is volatilized. The solution is then neutralized to methyl red, and finally titrated to phenolphthalein with standard sodium hydroxide.

Although boric acid is too weak to be titrated to phenolphthalein, advantage is taken of its reaction with polyhydric alcohols, such as mannitol, to produce stronger complex acids that can be titrated sharply. Hollander and Rieman [6] have studied the effect of various concentrations of mannitol on the sharpness of end point, and error in titration when boric acid is titrated with sodium hydroxide, and they conclude that a concentration of the order of 0.35 mole per liter (0.5 to 0.7 grams for each 10 ml of solution titrated) is optimum. The sodium hydroxide should be standardized against boric oxide as described by Hillebrand *et al.*[7]

PROCEDURE

1. Transfer a suitable weighed sample to a fused silica or platinum combustion cup, and ignite it in an oxygen bomb following the usual ASTM procedure[5] except that 5 ml of a 20% solution of Na_2CO_3 should be used to wet the inside of the bomb.

> NOTE: If a sample contained 4% boron, a one-gram portion would give a titration of about 35 ml of $N/10$ NaOH.

2. After ignition, vent the excess of oxygen carefully, transfer the alkaline solution to a 500-ml Erlenmeyer flask equipped with a ground-glass joint, and evaporate it to about 25 ml. Acidify the solution to methyl red with $3N$ HCl, adding about 5 ml in excess, attach a reflux condenser to the flask, and boil for about 15 minutes to expel carbon dioxide.

> NOTE: A reflux condenser is required because of the tendency of boric acid to steam-distill.

3. Cool the solution, neutralize it to methyl red with carbonate-free $1N$ NaOH and then make the solution just pink with $N/10$ HCl. Add 0.6 g of mannitol for each 10 ml of solution and titrate with standard $N/10$ NaOH to the phenolphthalein end point. Calculate the percentage of boron in the sample, using 10.82 as the equivalent weight, and subtracting a reagent blank.

> NOTE: The blank correction amounts to a milliliter or so of $N/10$ NaOH and should not be omitted.

References

1. M. C. K. JONES and R. L. HARDY, "Petroleum Ash Components and Their Effect on Refractories," *Ind. Eng. Chem.*, 44:2615 (1952).
2. J. A. DEAN and C. THOMPSON, "Flame Photometric Study of Boron," *Anal. Chem.*, 27:42 (1955).
3. D. F. BOLTZ, Ed., *Colorimetric Determination of Nonmetals*, p. 343, Interscience, New York (1958).

4. W. M. BURKE, "Boron Determination in Volatile Organic Compounds," *Ind. Eng. Chem., Anal. Ed.,* 13:50 (1941).
5. AM. SOC. TESTING MATERIALS, Philadelphia, Pa., *1955 Book of ASTM Standards,* Part 5, D129.
6. M. HOLLANDER and W. RIEMAN, III, "Titration of Boric Acid in the Presence of Mannitol," *Ind. Eng. Chem., Anal. Ed.,* 17:602 (1945).
7. W. F. HILLEBRAND, G. E. F. LUNDELL, H. A. BRIGHT, and J. I. HOFFMAN, *Applied Inorganic Analysis,* 2nd Ed., p. 755, John Wiley & Sons, New York (1953).

Chapter 7

CALCIUM

Calcium is one of the more prevalent oil-soluble metals in crude oil,[1] and it is also introduced from contamination by drilling muds and brine. Salts from the latter sources, however, are ordinarily eliminated in a preliminary desalting process. The work of Gunn and Powers [2] indicates that calcium may occur in a volatile, oil-soluble form because they found it at an average concentration of 5 ppm in distillate feed stocks for catalytic cracking. In general, calcium in petroleum is nonvolatile, and direct ashing is the most satisfactory method of preparing a sample for determining it.

As there are no very satisfactory colorimetric methods for the determination of small amounts of calcium, it is necessary to resort to spectrographic procedures for concentrations less than about 50 ppm. Gunn and Powers,[2] and Karchmer and Gunn [3] have given details for the spectrographic determination of calcium and several other elements in the range of 0.1 to 10 ppm. Flame photometry can also be applied with advantage in some special cases.

Calcium is one of the more important elements determined in the petroleum analytical laboratory. It was one of the earliest of the ash-type detergent additives used in lubricating oils, and although the present trend is toward ashless additives, calcium sulfonates and other salts are still widely used, and no doubt will continue to be for some time to come. Calcium naphthenate is a familiar paint drier, and calcium stearate and lime are regularly used in the manufacture of grease.

By far the most common forms of calcium additives for lubricating oil are the sulfonates, although others, such as

78

calcium phenate derivatives, are also employed. Calcium is rarely used alone in compounded oils; additional elements, e.g., phosphorus, sulfur, and zinc, may also be present. Barium is not ordinarily used with calcium in compounding oils, but in the procedure provided here provision is made for its presence.

Additives containing relatively high percentages of phosphorus, sulfur, and zinc are sometimes stabilized against contamination from water by adding a small amount of calcium salt. The determination of calcium in the presence of perhaps 250–300 times as much zinc and phosphorus presents some difficulties, one of the more serious being that some of these mixtures generate water as they burn, and spatter uncontrollably. Further disadvantages are that 1) the excessive amounts of salts produced in taking a large enough sample for the permanganimetric determination of calcium are difficult to handle by soft ashing and wet oxidation; and 2) direct ashing produces fused residues. Although this is not an especially important determination it can be readily accomplished by an extraction procedure followed by a versenate titration, as described in Section 2.

1. The Determination of Calcium in New Lubricating Oils and Additives

Because of the importance of the calcium additives, it is appropriate to examine the more significant considerations in the determination of the element. For the destruction of organic material, the best procedure is direct ashing, and for the final determination the oxalate-permanganate procedure leaves little to be desired. The recommendations of Kolthoff and Sandell [4] are followed in the precipitation of calcium oxalate: Calcium is precipitated by adding an excess of ammonium oxalate to the acid solution of the metal, and neutralizing with ammonium hydroxide, as this procedure yields the most satisfactory precipitate for permanganimetric determination. Other authors [5,6] have described variations in the oxalate precipitation to permit the determination of calcium in the presence of aluminum, iron, magnesium, manganese, phosphorus, silica, sodium, and titanium, using a single precipitation. This is accomplished by making the precipitation from an acid solution with a final pH of about 3.7.

The procedure given here may be used for the determination of calcium in naphthenate driers, additives, and compounded lubricating oils containing any or all of the following elements: barium, phosphorus, sulfur, and zinc. All of these products may be ashed directly in porcelain. In the exceptional case when fluxing agents such as sodium sulfonate are present, the direct ashing method is still applied, but the sample is pretreated with a few drops of fuming sulfuric acid, and the temperature for final ignition should not exceed 550°C. Ignited calcium sulfate dissolves readily in dilute hydrochloric acid, and even with as much as 3% of sodium sulfonate present the dish is not attacked (Table 7:1, Item IV).

The method of wet oxidation is often recommended for samples that yield fused ashes, but the procedure is undesirable. Sulfate is coprecipitated with calcium oxalate because of the low solubility of calcium sulfate, and this causes low results for the permanganate method. Low results are also obtained when barium is present (see Chapter 5, Table 5:1, and Item III, Table 7:1).

The effect of sulfur in the sample is ordinarily insignificant as much of it is volatilized during the initial burning, and a double precipitation eliminates the difficulty in any case. The coprecipitation of phosphate is not significant when precipitation is carried out in acid solution as in the procedure described here.

When determining calcium in the presence of a large quantity of barium in solution, double, or perhaps even triple precipitations may be required. If the barium has been laid down in the ash as barium sulfate by pretreatment with sulfuric acid, a single precipitation is sufficient, but results for calcium may be slightly low, probably because of mechanical inclusion. Items I and II in Table 7:1 are the results of single and double precipitations after direct ashing with, and without pretreatment with, a few drops of fuming sulfuric acid.

TABLE 7:1

DETERMINATION OF CALCIUM IN PRESENCE
OF BARIUM

Sample Prepared by *		Single Pptn.	Double Pptn.
I	Direct Ashing	0.223	0.196
II	Direct ashing with sulfuric acid pretreatment	0.191	0.190
III	Soft ashing, wet oxidation	0.185	0.186
IV	Same as II, with 3% sodium sulfonate added	0.197	....

* Sample contains 0.42% barium and 0.197% calcium; no other metals present.

Unless there is a preponderance of zinc calcium is easily determined in its presence; usually a single precipitation is sufficient. Thus, when 51.2 mg of calcium was precipitated as described in the procedure given here, in the presence of 50 mg of zinc, the average determination, based on four separate single precipitations, was 51.0 mg (0.4% low). (See Section 3 for a special procedure for determining small amounts of calcium in samples that are high in phosphorus, sulfur, and zinc.)

PROCEDURE

1. Weigh a suitable sample and transfer it to a Coors 2/0A porcelain dish.

NOTE: The amount of sample should be enough to yield at least 10 mg of calcium, and not more than 100 mg.

2. Ignite the sample of oil with a burner and allow it to burn freely until the fire goes out. Burn the loose carbon from the edge of the dish, and heat the residue with a light flame until a soft ash remains. Transfer the dish to a furnace, and ignite at 550°C until the carbon has been oxidized.

NOTES: The dish should not be heated strongly with a blast burner as mechanical losses or fusion of the ash may occur.
If fluxing agents are known to be absent temperatures much higher than 550°C may be used.

3. When the ash is clean, remove the dish from the furnace and allow it to cool to room temperature. To the residue add about 2 ml of HCl, and roll the dish to wet the entire ash. Add 10 ml of water and warm the dish on the steam plate for a few minutes. Remove it from the steam plate, add about 20 ml of water, and if there is insoluble material filter the solution into a 250-ml beaker through Whatman No. 42 paper, washing with water. Discard the residue.

NOTES: If the ash dissolves immediately there is no need for further heating. If there is an insoluble residue the heating should be continued for 15 minutes; or longer, if the residue is large.
The residue is most likely to be barium sulfate.

4. Dilute the solution to about 150 ml, add approximately 25 ml of a solution of ammonium oxalate containing 3 g of $(NH_4)_2C_2O_4 \cdot H_2O$, and a few drops of methyl red indicator. Heat the solution to about 80°C, neutralize by adding NH_4OH drop-by-drop until the indicator just turns yellow, and then add 1 ml of HAc. Allow the solution to stand without further heating, but with occasional stirring, for one hour, or until the supernatant liquid is clear.

NOTES: The solution must be hot to dissolve this much ammonium oxalate, it should be filtered if not clear.
The final pH is about 3.7.

5. Filter the solution through Whatman No. 40 paper, washing with water, and transferring as much as possible of the precipitate to the paper. Dissolve the precipitate in 25 ml of 1:1 HCl, receiving the solution in the original beaker, and washing the paper thoroughly with water. Dissolve about 1 gram of $(NH_4)_2C_2O_4 \cdot H_2O$ in a small volume of water, and add it to the acid solution. Dilute this to about 150 ml, and reprecipitate the calcium oxalate as described.

NOTE: With many samples a single precipitation is sufficient.

6. When the supernatant liquid is clear, filter the precipitate on Whatman No. 40 paper, washing with water. Wash off the stem of the funnel with water, and discard the filtrate. Place the original beaker under the funnel, fill

the paper half-full with water, punch a hole in the paper with a small stirring rod, and wash the precipitate into the beaker. Fold the washed paper over the side of the beaker, add 100 ml of water and 25 ml of 1:1 H_2SO_4, heat the solution to about 80°C, and titrate slowly with standard $0.1N$ $KMnO_4$ to a stable pink end point. Push the paper into the solution, stir to disintegrate it, and complete the titration. Calculate the percentage of calcium in the original sample, using an equivalent weight of 20.04.

NOTE: If the solution is cloudy after heating to 80°C it indicates that the separation from barium was not complete.

2. Determination of Calcium in Used Lubricating Oils

Calcium occurs in most used oils because it was used as compounding. The presence of calcium chloride has also been reported [7] as a result of anti-freeze solutions leaking from the cooling system, but this is relatively rare. If the history of the sample is unknown account must be taken of a number of metals (see Chapter V, Section 2); and if different types of compounded oils have been added to the crankcase a substantial number of elements may be present in significant amounts. The fact that these oils are often wet is still another difficulty.

For the determination of barium in used oils the direct ashing procedure was rejected because a carbonate fusion would have been necessary to open up the residue after ignition and the material of the dish (porcelain) precluded this. The case is otherwise with calcium, however, because it can be dissolved virtually completely by leaching the residue with dilute hydrochloric acid, the insoluble material being discarded.

By pretreating the weighed sample of oil with a few drops of fuming sulfuric acid, two of the more troublesome elements, barium and lead, are brought down in the ash as sulfates, and little of either passes into the hydrochloric acid solution. It cannot be assumed, however, that all of the barium is retained as the sulfate for it is likely that some barium sulfide will be formed by carbon reduction during ignition, and pass into the filtrate. Although the barium sulfide in the ash could be reconverted to the sulfate by treating the ash with sulfuric acid, this would render calcium sulfate insoluble (see Chapter 2, Section 2). Aluminum, chromium, and iron do not dissolve to any great extent during the leaching with dilute acid, and other elements are easily separated from calcium by a double precipitation.

When the wet oxidation procedure is applied to used oil that contains barium or lead, the insoluble sulfates of these metals carry with them significant amounts of calcium (see Chapter 5, Table 5:1) which can only be recovered by a time-consuming carbonate fusion. As mentioned before, a further disadvantage of wet oxidation, even in the absence of heavy metals, is the coprecipitation of calcium sulfate with calcium oxalate, which causes low results. This latter effect by itself is more or less corrected by a double precipitation. Item III in Table 7:1 illustrates the two effects combining to yield significantly low results for calcium.

To check the procedure given here a synthetic sample was prepared. The composition of the sample and the results of single and double precipitations of calcium are shown in Table 7:2.

TABLE 7:2

EFFECT OF SINGLE AND DOUBLE
PRECIPITATION OF CALCIUM

Aluminum	0.07%
Barium	0.37%
Calcium	0.20%
Iron	0.09%
Lead	0.18%
Zinc	0.17%
Calcium, single pptn.	0.212%
Calcium, double pptn.	0.202%

Before giving the details of the procedure, a few suggestions for handling different types of samples may be useful. A porcelain dish of suitable capacity should be used for the incineration and ignition of all samples. For additives, driers, and compounded oils, the sample is weighed, incinerated, and finally ignited at 550°C until all carbon is oxidized. As naphthenate driers are likely to spatter slightly, it is advisable to place a piece of filter paper over the surface of the sample before igniting it with a burner.

The pretreatment of certain samples with a few drops of fuming sulfuric acid has been discussed in Chapter 2, Section 2. This treatment should be applied to all used oils; to samples that are known to contain lead; to samples in which volatility of metals is a consideration (distillates of all kinds) ; and it is convenient to use if calcium is to be determined in light, powdery materials such as calcium stearate. As has been mentioned, the sulfuric acid pretreatment prevents the formation of fused ashes when fluxing agents such as alkali, or lead salts are present, and it is a valuable aid in handling such samples.

For paint driers (calcium naphthenate), and new lubricating oils that do not contain barium or lead, a single precipitation usually suffices. Double precipitations should always be made when barium or lead may be present, and when handling used lubricating oils.

PROCEDURE

1. Weigh a suitable sample and transfer to a Coors 2/0A porcelain dish. To the sample add 5 drops of fuming sulfuric acid, and mix thoroughly with a small glass rod. Heat the mixture for a few minutes on the steam plate with occasional stirring. Remove the rod, wipe it with a piece of filter paper, and drop the paper into the dish.

NOTE: The amount of sample should be enough to yield at least 10 mg of calcium, and not more than 100 mg.

For steps 2 to 6, follow the procedure for *New Lubricating Oils* . . ., noting the following observations:

2. A temperature of 550°C should be maintained for used oils, and for samples that may contain fluxing agents; for most compounded lubricating oils and driers much higher temperatures may be used.

3. In addition to $BaSO_4$, the residue may contain Al_2O_3, Cr_2O_3, etc.

3. Estimation of Smaller Concentrations of Calcium

The procedure that follows can be used for determining smaller amounts of calcium (.02–.05%) in the presence of a preponderance of phosphorus, sulfur, and zinc. Zinc thiophosphate derivatives used as additives for lubricating oil are sometimes stabilized by adding calcium salts, and the determination of the latter is not practical by the procedures given in Sections 1 or 2. Some of these derivatives generate water, or perhaps phosphoric acid, on burning, and spatter violently. The expedient of placing a piece of filter paper in the sample to serve as a wick is not very effective with these products; and even if the sample could be burned without mechanical loss there would be too much zinc phosphate to separate by the oxalate procedure. Moreover, direct ashing causes the formation of phosphate glasses which attack porcelain, enclose carbon—thus preventing complete ignition—and cannot be effectively dissolved with hydrochloric acid.

Mr. Leo Heller of the California Chemical Company has contributed the following procedure wherein calcium and zinc are extracted with dilute acid from a sample dissolved in a mixed solvent. Zinc is then complexed with cyanide, and calcium is titrated with standard Versenate solution using murexide as indicator. A Versenate procedure has been described by Gerhardt and Hartman [8] for determining either calcium or zinc in additives and oils, but only one bivalent metal may be present. Bertolacini [9] has developed rapid routine control methods for barium, calcium, and zinc using Versenate complexometric titrations after extraction with hot hydrochloric acid.

As this application is not particularly important, the procedure here is designed to give merely an estimation (±10%) of small amounts of calcium present with large amounts of phosphorus and zinc.

PREPARATION

1. REAGENTS:

  Alkali-Cyanide Solution: Dissolve 10 grams of sodium hydroxide and 10 grams of potassium cyanide in 100 ml of water.

Extraction Solution: Mix 4 ml of concentrated hydrochloric acid and 70 ml of *n*-butyl alcohol with 900 ml of water.

Murexide Indicator: Dissolve 0.1 gram of murexide in 50 ml of water. Prepare fresh weekly.

Standard Calcium Solution: Slurry 0.50 grams of calcium carbonate with water, dissolve by adding drops of hydrochloric acid, and dilute to 500 ml with water. 1 ml = 1 mg $CaCO_3$.

Standard Versenate Solution: This may be purchased, or prepared by dissolving about 4 grams of disodium ethylenediamine tetracetate in one liter of water, and adjusting to equivalence with the standard calcium solution.

PROCEDURE

1. Weigh 2.5 to 3.5 g of sample, dissolve it in approximately 50 ml of benzene, and transfer it to a separatory funnel containing 25 ml of extraction solution. Shake the mixture for 30 seconds, and allow it to stand until the layers separate completely. Transfer the lower layer to another funnel containing 25 ml of benzene, and shake the mixture for 30 seconds. To the first funnel, add 25 ml of extraction solution, and shake the mixture. When the lower layer in the second funnel has separated, drain it into an Erlenmeyer flask, transfer the lower layer in the first funnel to the second funnel, and shake the latter again.

2. Using a graduate, measure 5 ml of the alkali-cyanide solution into the flask, and add 5–10 drops of the murexide indicator. Titrate slowly with standard Versenate to an orchid tint. Add the lower layer in the second funnel to the flask, and complete the titration.

NOTES: The usual precautions should be followed in handling the cyanide solution.

If in doubt as to the end point, back-titrate with standard calcium solution to a red color, and return to the orchid shade with Versenate solution. The solutions are equivalent, and the titration with calcium solution can be subtracted from the total

Versenate titration to give the "sample titration" (see Note under 3).

3. Prepare a blank solution by shaking 50 ml of the extraction solution with 75 ml of benzene. Titrate the lower layer of this single extraction, after adding alkali-cyanide and murexide, as in Step 2, and calculate the percentage of calcium in the sample.

NOTE: The method gives results for calcium accurate to ±10% of the amount present.

$$\% \ Ca = \frac{(sample \ titration - blank \ titration) \times 0.04}{weight \ of \ sample}$$

References

1. M. C. K. JONES and R. L. HARDY, "Petroleum Ash Components and Their Effect on Refractories," *Ind. Eng. Chem.*, 44:2615 (1952).
2. E. L. GUNN and J. M. POWERS, "Ash Residues from Petroleum Catalytic Cracking Feed Stocks," *Anal. Chem.*, 24:742 (1952).
3. J. H. KARCHMER and E. L. GUNN, "Determination of Trace Metals in Petroleum Fractions," *Ibid.*, 24:1733 (1952).
4. I. M. KOLTHOFF and E. B. SANDELL, *Textbook of Quantitative Inorganic Analysis*, 3rd Ed., p. 339, Macmillan (1952).
5. W. H. McComas, JR. and W. RIEMAN III, "Accurate Determination of Calcium Without Reprecipitation," *Ind. Eng. Chem., Anal Ed.*, 14:929 (1942).
6. J. J. LINGANE, "Volumetric Determination of Calcium," *Ind. Eng. Chem., Anal. Ed.*, 17:39 (1945).
7. H. LEVIN, K. UHRIG, and E. STEHR, "Radiator Antifreeze Materials," *Ibid.*, 11:134 (1939).
8. P. B. GERHARDT and E. R. HARTMAN, "Determination of Calcium or Zinc Additives in Lubricating Oils and Concentrates," *Anal. Chem.*, 29:1223 (1957).
9. R. J. BERTOLACINI, "Analyze for Ba, Ca, and Zn This Way," *Petroleum Refinery*, 37, No. 2:147 (1958).

Chapter 8

CHROMIUM

Chromium is a rather common oil-soluble contaminant of crude oils although its concentration seldom exceeds one- or two-tenths of a part per million. It is not particularly volatile, ordinarily being found only in the deepest gas oil cuts, and its concentration in these is usually less than .05 ppm. The element may be introduced as a contaminant into some petrochemical products from corrosion and heat-resistant steels, but the amount rarely exceeds 10 ppm. Chromium may occur as a metal from wear in used lubricating oils, and it is especially common in used lubricating oils from diesel locomotives.

As there is seldom any occasion to determine the element in such amounts that a volumetric procedure is required, none is provided. If there should be such a requirement, however, the sample can be prepared in the usual way by wet oxidation, with the addition of about two milliliters of 85% phosphoric acid. (Perchloric acid is not advisable in this instance, as will be shown later in the discussion.) The final determination can be made by the ferrous sulfate-permanganate method after oxidation with persulfate.[1]

Several metals cause cracking catalysts to lose selectivity, copper and nickel showing this property to the greatest extent.[2] In this respect chromium is about one-tenth as effective as these elements and it is not ordinarily of much significance because of its low concentration in distillates. Frequently it must be determined in the latter, however, and it is appropriate to consider the conditions for a precise determination in these stocks.

It may be presumed that the chromium content of a heavy gas oil will not exceed 0.1 to 0.2 ppm, and it usually will be less. To concentrate enough of the element for a colorimetric determination it is necessary to use a 100–200 g sample. Chromium has little tendency to volatilize during incineration and it can be recovered, essentially completely, by direct ashing. The insolubility of ignited chromic oxide, however, makes that process inconvenient, and the soft ashing, wet oxidation method is used in the procedure here. The pretreatment with sulfuric acid which is prescribed for some metals is without effect in the determination of chromium and is therefore unnecessary.

A considerable amount of tarry material and coke is produced in the incineration of heavy distillates, which is extremely difficult to oxidize. It may cause the preliminary fuming with sulfuric acid, as well as the subsequent oxidation with nitric acid, to be greatly prolonged, and lead to the separation of the insoluble acid sulfate, $Cr_4H_2(SO_4)7$. In the presence of phosphoric acid this salt does not separate from fuming sulfuric acid, and for this reason a milliliter or two of phosphoric acid is added before starting the oxidation.

An investigation by Hoffman and Lundell [3] indicates that small losses of chromium as chromyl chloride (CrO_2Cl_2) occur during oxidation with perchloric acid. Thus, when 100 mg of chromium is fumed with perchloric acid, losses of 0.2 to 0.4 mg occur, probably because of the formation of hydrochloric acid as a reduction product during the oxidation of trivalent chomium. By using perchloric acid to finish the oxidation in the procedure given here, recovery in the range of 0.01 to 0.05 mg is complete, whether or not phosphoric acid is present in the sulfuric acid solution. If larger amounts of chromium are to be handled, however, perchloric acid should not be used.

1. Colorimetric Determination of Chromium

As the amount of chromium to be determined is usually small a sensitive colorimetric procedure is required. The use of 1,5-diphenylcarbohydrazide (s-diphenylcarbazide) for the determination of chromium is well-established, and a comprehensive discussion of this reagent is given by

Sandell.[4] The stability of the reagent has been investigated by Urone,[5] who notes that it discolors and loses sensitivity in basic solvents such as water, methanol, ethanol, and the like. This is presumably caused by oxidation of carbazide to carbazone, a reaction which may be base-catalyzed. For the preparation of a solution that remains stable for a week or more, he recommends solvents such as methyl ethyl ketone; ethyl acetate solutions are stable for several months. Ege and Silverman [6] have reported that phthalic anhydride added to a solution of the reagent in ethanol will greatly improve stability, probably by maintaining an acidic environment. It is important to realize that once the reagent loses its sensitivity the loss cannot be compensated by increasing the excess, so if the solution shows any color it should be discarded.[5]

Interfering elements that should be considered include iron, molybdenum, and vanadium:

Iron produces a yellow color which is weakened in sulfuric acid, and is negligible in the presence of phosphoric acid. In any event it absorbs but little at 540 mμ, the wavelength at which the color of chromium is measured.

Molybdenum produces a red-violet color with the reagent which is much less intense than that with chromium, and the amounts that occur in distillates or crude oils (never more than 50 ppm in the most extreme examples), do not cause any difficulty.

Vanadium produces with the reagent a strong yellow-brown coloration, which is unstable and fades rather rapidly at the acidity specified in the procedure given. In many samples the presence of vanadium may be ignored, but to make the procedure generally applicable—even in such extreme cases as Venezuelan crude oils, in which vanadium may exceed chromium by a thousand-fold—a cupferron-chloroform extraction is provided which removes all three of the interfering elements under consideration, as well as copper, titanium, and others.

In the procedure under discussion, at the completion of the wet oxidation, the solution of the sample comprises about 5 ml of sulfuric acid, 1 ml of phosphoric acid, and somewhat less than 1 ml of perchloric acid. This is diluted with water to a volume of approximately 60 ml, producing an acidity appropriate for the chloroform extraction of the cupferrates of copper, iron, molybdenum, and vanadium. The extracts are discarded, and chromium remains in the acid solution free from interfering elements; but it is necessary at this point to introduce a step of oxidation to ensure the hexavalency of all of the chromium.

It is generally accepted that if the acidity of sulfuric acid is greater than 2N the oxidation of chromium by persulfate is inhibited. In the presence of phosphoric acid, however, the oxidation is complete, even when 5 ml of sulfuric acid and 1 ml of phosphoric acid are present in a volume of 50–60 ml. An acidity of 0.2 N is usually recommended for the color development with 1,5-diphenylcarbohydrazide because this reagent is unstable at higher acidities. Using the reagent prepared with phthalic anhydride,[6] however, the color is stable for 20 to 30 minutes in the very strongly acidic solution under consideration. The reason for this has not been ascertained but it makes possible the direct determination of chromium without further operations.

The preceding discussion has been concerned with the determination of chromium in distillates and crude oils, but the same general procedure is applicable for other materials such as petrochemicals that may have been contaminated by chromium alloys during manufacturing and processing. As mentioned in Chapter 2, organic carboxylic acids and anhydrides (e.g., benzoic acid, phthalic anhydride), are smoothly oxidized by 30% hydrogen peroxide. It is undesirable to apply the soft ashing, wet oxidation procedure to these materials because their manner of burning leads to low results. The step of extraction may be omitted because vanadium is not ordinarily present in these products unless vanadium pentoxide catalysts are involved in manufacturing them.

PREPARATION

1. REAGENTS:

   Cupferron Solution: Dissolve 6 grams of the ammonium salt of N-nitroso-N-phenylhydroxylamine

in 100 ml of water. Prepare freshly, immediately
before use.

Diphenylcarbazide Reagent: Dissolve 4.0g of phthalic
anhydride and 0.25 of 1,5-diphenylcarbohydrazide
in 100 ml of absolute ethanol. Discard when the
solution develops any color.

Chromium Standard: Dissolve .2828g of $K_2Cr_2O_7$ in
water and dilute to 1 liter. 1 ml = 0.10 mg Cr.

2. STANDARDIZATION:

Dilute the prepared chromium standard solution tenfold
to produce a solution containing .01 mg Cr/ml. Transfer
1.0, 2.0, 3.0, 4.0, and 5.0 ml of the dilute standard to 250-ml
beakers, and include a blank. To each beaker add 5 ml of
H_2SO_4 and 1 ml of 85% H_3PO_4, and dilute to about 50 ml
with water. Add 10 ml of N/10 $AgNO_3$ and 2 grams of
$(NH_4)_2S_2O_8$, heat to boiling, and boil 15 minutes.

Cool the solutions and transfer them to 100-ml volumetric
flasks, diluting in the process to about 85 ml. To each flask
add 5 ml of diphenylcarbazide reagent, dilute to volume, and
mix thoroughly. Measure the transmittances of the solutions
at 540 mμ within 20 minutes and plot the results against the
corresponding milligrams of chromium on semilogarithmic
graph paper.

PROCEDURE

1. Weigh and transfer a well-mixed sample to a clean,
dry 400-ml beaker, place the beaker in a No. 2 tin can, and
ignite the oil with a flame. When the sample is reduced to
a soft ash, burn the carbon from the walls of the beaker
as described in Chapter 2, Section 2.

NOTE: The sample should be sufficient to contain 0.01–0.05 mg
of chromium. For light gas oils 300 grams can be handled conveniently. The deeper gas-oil cuts, especially from waxy crudes,
yield large amounts of tarry material, and 100 grams is the practical limit. For most crudes 50 grams may be taken.

2. Cool the beaker, add about 20 ml of sulfuric acid, heat
to fumes, and digest the mixture on the hot plate until the
material is thoroughly wet with the acid. Add about 1 ml
of 85% H_3PO_4, and oxidize the organic material with nitric

acid as described in Chapter 2, Section 3. When a clear solution has been obtained, cool it, add 5 ml of HNO_3 and 1 ml of $HClO_4$, and heat the solution until the perchloric acid boils.

NOTES: If the residue is large, as much as 40 ml of sulfuric acid may be required.
The length of time of the digestion depends upon the sample. In extreme cases an hour may be required; the sulfuric acid must be replenished from time to time.

3. Before all of the perchloric acid has been volatilized remove the beaker from the heat and cool it to room temperature. Dilute the contents with about 25 ml of water, heat it to boiling, and boil gently for about 5 minutes.

NOTE: In the majority of instances a clear solution is obtained; boiling is required to dissolve aluminum, iron, nickel, etc.

4. If the presence of vanadium was indicated during the oxidation, transfer the cooled solution to a 125-ml separatory funnel, diluting it in the process to 50–60 ml. (If vanadium is not present, dilute to 50–60 ml and proceed with the persulfate oxidation.)

NOTE: The presence of vanadium is indicated by a series of color changes which can be observed when the oxidation is near completion. The color in fuming sulfuric acid is green; when nitric acid is added the orange color of vanadate is observed; when the nitric acid is evaporated the pentavalent compound decomposes and the color reverts to green. The addition of peroxide to vanadate solution produces a transient brown coloration (pervanadic acid) followed by the return of green. Boiling perchloric acid produces a deep orange color, or with smaller amounts, a bright yellow.

5. Add 2.5 ml of a freshly-prepared 6% solution of cupferron, and mix. Extract with a 10-ml portion of chloroform, shaking for 30 seconds. Draw off and discard the chloroform layer, and extract the other layer with an additional 5-ml portion. Add 0.5 ml of cupferron solution, and extract successively with one 10-ml, and two 5-ml portions of chloroform, discarding all of the chloroform extracts.

NOTES: If the determination of copper, iron, molybdenum, or vanadium is required, the chloroform extracts should be combined, and reserved.

If several milligrams of iron or vanadium are present, more cupferron may be required than is specified here.

6. Drain the acid layer into a 250-ml beaker and boil it until the residual chloroform is evaporated. To the boiling solution add 10 ml of $N/10$ AgNO$_3$ and 2 grams of (NH$_4$)$_2$S$_2$O$_8$, and continue boiling for 15 minutes.

NOTES: Persulfate will react with chloroform to produce chloride which in turn precipitates the catalyst (Ag+). The solvent is therefore evaporated as completely as possible at this point.
This salt is best added dissolved in a few milliliters of water.

7. Cool the solution and transfer it to a 100-ml volumetric flask, diluting it in the process to 85 ml. Add 5 ml of diphenylcarbazide reagent, dilute to volume, and mix. Measure the transmittance of the solution at 540 mμ within 20 minutes, and from the prepared calibration curve read the corresponding mg of chromium.

NOTES: There may be a little AgCl in the solution. If so, it should be removed by filtration, receiving the filtrate in the volumetric flask.
If the cupferron extraction has been omitted, there may be a brown color developed at this point from small amounts of vanadium; it will disappear in a few minutes and cause no difficulty.
In the preparation of the calibration curve, the milliliters of dilute standard chromium specified correspond to .01, .02, .03, .04, and .05 mg of chromium.

References

1. W. F. HILLEBRAND, et al., Applied Inorganic Analysis, 2nd Ed., p. 527, John Wiley and Sons, New York (1953).
2. G. A. MILLS, "Aging of Cracking Catalysts," Ind. Eng. Chem., 42: 182 (1950).
3. J. I. HOFFMAN and G. E. F. LUNDELL, "Volatilization of Metallic Compounds from Solution in Perchloric or Sulfuric Acid," J. Research Nat'l. Bur. Standards, 22:465 (1939).
4. E. B. SANDELL, Colorimetric Determination of Traces of Metals, 2nd Ed., p. 260, Interscience, New York (1950).
5. P. F. URONE, "Stability of Colorimetric Reagent for Chromium, s-Diphenylcarbazide, in Various Solvents," Anal. Chem., 27:1354 (1955).
6. J. F. EGE, JR., and L. SILVERMAN, "Stable Colorimetric Reagent for Chromium," Ind. Eng. Chem., Anal. Ed., 19:693 (1947).

Chapter 9

COBALT

Cobalt occurs in crude oils only in isolated samples and in trace amounts. It has been found, for example, in two samples of Louisiana crude oil at the .01 ppm level; in most crudes it is not detectable. The element would probably alter the selectivity of cracking catalysts with the same order of effectiveness as copper and nickel, but it is never found in distillates, and is not a factor in selecting feed stocks for catalytic cracking.

As there is rarely any occasion to determine small amounts of the element in petroleum, no procedure will be given here. Attention is directed, however, to a thiocyanate procedure described by Kitson [1] for determining moderate amounts (0.05 to 0.5 mg) of cobalt. If a more sensitive method is required, the Nitroso-R Salt procedure as described by Sandell [2], is recommended. Volatility may or may not be a consideration in preparing the sample; when in doubt, the sulfuric acid pretreatment followed by soft ashing and wet oxidation should be used. (See Chapter 2, Section 2.)

Cobalt molybdates are used as catalysts in cyclization, dehydrogenation, and isomerization. Although these materials are often analyzed in the petroleum analytical laboratory, the details are beyond the scope of this book. It may be noted, however, that the volumetric ferricyanide procedure given here is applicable only if care is taken to pretreat the ground sample of catalyst with hydrogen peroxide; high results are obtained if any of the molybdenum is in a reduced state.

As the determination of cobalt would probably be most commonly required for the manufacturing control of cobalt naphthenate driers for paints and inks, this procedure will be discussed in detail following the two general methods given here.

1. Electrolytic Determination of Cobalt

PROCEDURE

1. Transfer a 100-ml aliquot of the prepared standard cobalt solution to a 180-ml electrolytic beaker. Add 20 grams of $(NH_4)_2SO_4$, and 40 ml of NH_4OH, place the beaker on the electrolysis stand, immerse a tared platinum gauze cathode and a rotating platinum anode in the solution, and electrolyze at a potential of 6 volts and a current of 0.5 to 1.0 amperes for one hour.

NOTES: The preparation of the standard solution is described in Section 2 under *Reagents*.

If this procedure is applied to the solution of a sample, excess acid must be neutralized with ammonium hydroxide until a permanent precipitate is obtained. The specified amounts of ammonium sulfate and excess ammonium hydroxide are then added, followed by one gram of sodium sulfite.

The completion of the electrolysis may be checked by withdrawing 5 ml of the electrolyte and evaporating in a porcelain crucible with a few crystals of phenylthiohydantoic acid; the formation of a pink or reddish color indicates the presence of cobalt.

2. When deposition is complete, lower the beaker from the electrodes with the current flowing. Wash the cathode first with water, then with alcohol, finally drying it briefly in an oven. Re-weigh the cathode and deposit, and compute the titer of the cobalt standard solution.

NOTE: It is convenient to express the concentration of this solution as mgCo $++$/ml.

2. Volumetric Determination of Cobalt

The oxidation of cobaltous ion by ferricyanide in ammoniacal ammonium citrate solution was first proposed as a volumetric method for the determination of cobalt by Dickens and Maassen.[4] Later, Steele and Phelan [5] applied the method to steels and alloys.

Although the method evidently has not found general

acceptance, it is remarkably free from interferences, and it gives precise results. Aluminum, chromium (VI), copper, iron, lead, molybdenum (VI), nickel, vanadium (V), and zinc are without effect; manganese is the only common metal that interferes, and if it is present it must be removed.

The determination is made by pouring the prepared solution of the sample into a measured excess of ferricyanide in a supporting electrolyte of ammonium citrate and ammonium hydroxide. The excess ferricyanide is titrated with standard cobalt solution, the reaction being followed potentiometrically using a platinum electrode and a calomel electrode equipped with a half-saturated sodium sulfate salt bridge. If a vacuum tube voltmeter, or a commercial pH meter with a voltage scale is used, the inflection occurs between $+100$ and 0 millivolts.

As remarked before, the use of perchloric acid will indicate the presence of manganese. If necessary, it can be removed by boiling the diluted acid solution with ammonium persulfate, filtering, and discarding the resulting manganese dioxide. Excess persulfate must be destroyed before continuing with the titration.

PREPARATION

1. REAGENTS:

> *Ammonium Citrate Electrolyte:* Dissolve 250 grams of citric acid in water, slowly add 350 ml of ammonium hydroxide, cool, and dilute to one liter with water.

> *Potassium ferricyanide solution:* Dissolve 11 grams of $K_3Fe(CN)_6$ in water, and dilute to one liter. As this solution is light-sensitive it should be stored in a brown bottle.

> *Standard cobalt sulfate solution:* Dissolve 9.5 grams of $CoSO_4 \cdot 7H_2O$ in water and dilute to one liter. Standardize by the electrolytic procedure, Section 1.

2. STANDARDIZATION OF COBALT SULFATE SOLUTION:

For the preparation of the standard solution required here the electrolytic procedure is reliable and convenient. As reagent grade cobalt sulfate containing 0.01 per cent nickel, or less, is readily available, a correction for nickel is unnecessary. Manganese is present in negligible amounts, and as the electrolysis may be carried out without the addition of sodium sulfite there is no need of a correction for sulfur. The amount of platinum transferred to the cathode (0.3 to 0.5 mg) is negligible in a 200-mg deposit, and in fact, it more or less compensates for a similar amount of cobalt that is left in solution. It is recommended that cobalt sulfate rather than the nitrate be used to prepare the standard solution because the nitrate sometimes produces a spongy, dark deposit in the electrolytic procedure, and because cobalt sulfate solutions are more stable in storage.

PROCEDURE

1. Transfer a suitable weighed sample to a 400-ml beaker, cover with a ribbed watch glass, add 10 ml of H_2SO_4 and 5 ml of HNO_3, and allow the initial vigorous reaction to subside. Heat the mixture cautiously at first, gradually raising the temperature until sulfuric acid fumes are evolved. Continue the wet oxidation in the usual way (Chapter 2, Section 3) and when the solution is clean add 5 ml of HNO_3 and 2 ml of $HClO_4$. Heat until the HNO_3 is evaporated and the $HClO_4$ begins to boil. If manganese is present, a brown color or precipitate will be formed.

NOTES: As these samples are volatile they must be weighed in a Lunge weighing bottle or gelatine capsule. A size 000 capsule will contain about one gram of cobalt naphthenate which is an appropriate quantity (50–70 mg Co).

There is usually an induction period before this reaction starts. If a capsule is used it should be opened when it is transferred to the beaker, or else disintegrated in sulfuric acid before any nitric acid is added.

Manganese is ordinarily not present in commercial cobalt naphthenates; a very slight brown color may be disregarded. In the subsequent titration 1 mg of Mn is equivalent to 1.07 mg Co.

2. Evaporate the solution until the volume of sulfuric acid is reduced to about 2 ml, and cool. If manganese is to be removed, dilute to about 100 ml, add 2 grams of $(NH_4)_2S_2O_8$, heat to boiling, and boil 10 minutes. Cool the solution and filter through Whatman No. 40 filter paper, washing with water. Discard the paper and precipitate and

evaporate the filtrate to fumes of sulfuric acid. Cool, and dilute to about 50 ml with water.

NOTE: In the absence of silver ion the decomposition of persulfate is slow, and to be sure that it is destroyed it is best to evaporate to fumes. If any persulfate is present when the titration is made, low results are obtained.

3. To a clean beaker add 50 ml of the ammonium citrate solution and 100 ml of concentrated ammonium hydroxide. With a pipette, transfer 50.0 ml of potassium ferricyanide solution to the mixture and mix well. Pour the cool sample solution into this mixture, place the electrodes (platinum and calomel with salt bridge) in the solution, and titrate potentiometrically with standard cobalt solution, recording the titration.

NOTES: The same pipette should be used for both the sample and the blank. A new blank must be determined each day as the ferricyanide solution is slightly unstable.

To avoid the formation of troublesome precipitates the cobalt solution must be poured into the ammoniacal ferricyanide solution rather than the reverse.

4. Prepare a similar solution as a "blank", titrating with the standard cobalt solution as before. The difference between the "blank" titration and the sample titration multiplied by the titer of the standard cobalt solution gives the milligrams of cobalt in the sample.

DETERMINATION OF COBALT IN NAPHTHENATE DRIERS

The manufacture of cobalt naphthenate comprises the following steps: Metallic cobalt is dissolved in sulfuric acid, and combined in a mixer with thinner and a mixture of segregated sodium naphthenates of specified acid number and color; the mixture is washed with water until excess soap is removed; it is dehydrated and brightened by a few hours heating, during which some thinner is also evaporated; the amount of cobalt is then determined, and an appropriate amount of thinner is added to produce the desired content of metal.

When different metallic driers are prepared in the same mixer there is a possibility of cross-contamination, but this is not considered much of a manufacturing problem, because the mixer and associated equipment are washed thoroughly between batches, and small amounts of foreign metals (about 0.1%) ordinarily have no effect on the action of a drier. In developing an analytical procedure, however, it is appropriate to consider the effects of possible contaminants, and to provide for their presence if necessary. Commercial metallic cobalt contains from 0.2 to 0.5 per cent nickel, and 0.1 to 0.3 per cent iron; the manganese content is usually negligible. Other common driers that may be manufactured in the same equipment include calcium, copper, iron, lead, manganese, and zinc. The effects of these metals, and others, will be considered in the discussions of the two procedures provided here.

For reasons mentioned in Chapter 2, Section 1, direct ashing is unsatisfactory for recovering cobalt from petroleum products. The method of pyrosulfate fusion described in Section 4 of the same chapter can be used, but it offers no special advantage, and for the analysis of driers the direct wet oxidation of the sample in sulfuric acid is the most convenient method for destroying the organic material. In Chapter 2, Section 3, it was pointed out that heavy materials dissolved in thinners are likely to spatter if heated with sulfuric acid alone. For this reason, nitric acid should be added to the sample dispersed in sulfuric acid, and the initial vigorous reaction should be allowed to subside before heating is begun.

Because the solvent is extremely volatile, the sample must be weighed in a Lunge weighing bottle or, if desired, in a gelatine capsule. The capsule and sample are dropped into a beaker, and sulfuric acid is added and swirled until the capsule disintegrates, after which the nitric acid treatment is begun. The use of perchloric acid to finish the oxidation is desirable as it will indicate the presence of manganese— an element that interferes in the volumetric procedure. Oxidizing it to manganese dioxide by perchloric acid easily reveals the presence of as little as one milligram of manganese in the solution.

Although the electrolytic method is often recommended for the determination of cobalt, it is not very convenient for the analysis of cobalt naphthenate driers. Several possible contaminants, namely copper, manganese, nickel, and zinc, are deposited on the cathode with cobalt, and although the deposition of manganese can be prevented by the addition of sodium sulfite, the latter introduces sulfur into the deposit. A slight dissolution of the anode transfers a small amount of platinum to the cathode, the amount increasing with the time of the electrolysis. In addition, ferrous iron, if present, has a disturbing effect. All of these interferences tend to produce high results, and corrections for nickel and sulfur, as described by Hillebrand et al.,[3] are especially important if accurate results are to be obtained. Electrolytic results are invariably a little higher than those by the volumetric procedure. Because of these difficulties the electrolytic procedure is not recommended for the determination of cobalt in naphthenate driers.

References

1. R. E. KITSON, "Simultaneous Spectrophotometric Determination of Cobalt, Copper, and Iron," Anal. Chem., 22:664 (1950).
2. E. B. SANDELL, Colorimetric Determination of Traces of Metals, 2nd Ed., p. 274, Interscience, New York (1950).
3. W. F. HILLEBRAND, G. E. F. LUNDELL, H. A. BRIGHT, and J. I. HOFFMAN, Applied Inorganic Analysis, 2nd Ed., pp. 412–414, John Wiley & Sons, New York (1953).
4. P. DICKENS and G. MAASSEN, Archiv Eisenhüttenw, 9:487 (1936).
5. G. J. STEELE and J. J. PHELAN, "The Determination of Cobalt in Steels and Alloys," Gen. Elec. Rev., 42:218 (1939).

Chapter 10

COPPER

Soluble copper occurs in a number of crude oils to the extent of about one-tenth of one per cent of the ash.[1] The element is often found in deposits resulting from corrosion of such alloys as monel, but it is rarely introduced from this source in a petroleum-soluble form. It is an undesirable element in gasolines, and this determination is covered in Section 1. Copper naphthenate is commonly used as a paint drier, as a water and rot-proofing agent for fabrics, and in wood preservatives. Other copper organic salts are occasionally used as inhibitors of corrosion and oxidation; the procedure in Section 2 is applicable to these products.

Copper ranks with nickel in detrimentally altering the selectivity of cracking catalysts, and it is frequently necessary to determine it in distillate feed stocks; the details of this application are covered in Section 3. Finally, the examination of used lubricating oils often reveals the presence of more or less copper. If present as a metal from wear its concentration is small; if the additives against oxidation and corrosion, mentioned above, are present the amounts will be significant. Section 4 describes the determination of copper in used lubricating oils.

Table 2:1 in Chapter 2 indicates that five of the ten methods described for preparing a petroleum sample for inorganic analysis may be applied for the determination of copper. Section 1 here, for handling gasolines, prescribes a dilute acid extraction; Section 2, for naphthenate driers, additives, wood preservatives, etc., uses direct wet oxidation, with pyrosulfate fusion as an optional alternative; Section 3, copper in distillates, uses the soft ashing, wet oxidation method (following a pretreatment with fuming sulfuric acid) ; and in Section 4, copper is recovered from used oils by treating with alkaline sulfide.

Before considering the details of these four procedures, attention is called to some special methods that may be of interest. Buchwald and Wood [2] have applied an ion-exchange technique for recovering copper from mineral oils. The sample is percolated through a cation-exchange column, copper is then eluted with dilute sulfuric acid, and determined colorimetrically. Zall, et al.,[3] determine copper in diesel fuel oil colorimetrically by adding neocuproine (2,9-dimethyl-1,10-phenanthroline) directly to the sample of oil. No ashing is required; chloroform and isopropyl alcohol are used as solvents. Spectrographic methods have been given by Barney [4] for copper in turbine oils, by Hodgkins and Hansen [5] for copper in crankcase oils, and by Karchmer and Gunn [6] for copper in distillates used in feed stocks for catalytic cracking.

In Sections 1 and 3, colorimetric procedures using sodium diethyldithiocarbamate are prescribed. The yellow copper salt is extracted from ammoniacal citrate solution with carbon tetrachloride, and the transmittance of the solution is measured spectrophotometrically. Interferences and other details are discussed in these sections.

For determining larger amounts of copper, as in Sections 2 and 4, the iodometric procedure has been selected. The empirical standardization of thiosulfate against metallic copper was formerly advocated, but the work of Foot and Vance,[7] Crowell,[8] and Hammock and Swift,[9] has shown this to be unnecessary if potassium thiocyanate is added just before the end point. The procedure is so well-known that no further discussion will be given here. Kolthoff et al.,[10] have discussed the iodometric method in detail, and reference should be made to their book for further information.

The choice of the wet oxidation and iodometric procedure for determining copper in copper naphthenate in preference to the wet oxidation and electroplating method was influenced by the greater rapidity of the former. The method of wet oxidation and plating requires about two hours of working time with six to six and one-half hours elapsed time, whereas the procedure in Section 2 requires only about one hour. Speed is an important consideration in manufacturing control. In addition, an extensive investigation of the routine application of the two methods has indicated that the precision of the iodometric procedure is greater than that of the plating method, and it is also less subject to interferences.

Duplicate copper determinations by the two methods in 25 different samples of copper naphthenate, ranging from three to seven per cent copper, showed the following: the average difference between duplicates by the iodometric method was 0.016%; the maximum difference in 25 sets of duplicates was 0.06%; the average difference between duplicates by the electroplating method was 0.065%; the maximum difference of a single pair was 0.13%. In comparing the average results of duplicates by the two methods, it was noted that the plating method was usually higher. Specifically, the plating method averaged 0.03% higher with 14 samples, .04% lower with 7 samples, and gave the same average result with 4 samples.

The erratic results by the plating procedure are probably explained by the presence of nitrous acid in the solution after wet oxidation. Perhaps the addition of sulfamic acid to the solution before electrolysis, as recommended by Silverman,[11] would improve the results. Sulfamic acid reduces nitrous acid to nitrogen, and being more rapid, it is superior to urea for this purpose. Also, the acidity of the electrolyte is not decreased by the reaction

$$NH_2SO_3H + HNO_2 = N_2 + 2H^+ + SO_4^= + H_2O$$

1. Determination of Copper in Gasoline

Copper is an undesirable element in motor fuels and aviation gasolines because it induces the formation of gum. It is introduced in gasoline-soluble form in copper sweetening processes in concentrations of 0.01 to 0.2 parts per million. The feed stock for a sweetening process is ordinarily rectified straight-run gasoline, and the presence of small amounts of aliphatic acids probably accounts for the introduction of soluble copper.

To render the copper innocuous a metal deactivator is added to the sweetener product to a concentration of about 10 parts per million. There are several of these compounds commercially available, one of which will be mentioned for illustration. Tenemene 60 (Eastman Chemical Products, Inc.) is the trade name for a formulation consisting of 80% disalicylalpropylenediimine and 20% toluene.[12] The diimine forms a chelate with copper ion, inhibiting its undesirable gum-inducing properties.

Although the solubility of anhydrous inorganic copper salts in gasoline is extremely small, about 0.01 ppm, salts from the sweetening reagent may be carried mechanically in the gasoline stream. The subsequent addition of metal deactivator solubilizes these inorganic salts, thus increasing the copper content of the gasoline. The concentration of copper in sweetened gasoline is therefore usually higher after addition of deactivator than it is before, and the difference of these two determinations constitutes a rough measure of mechanical carry-over.

Thus, when approximately 50 mg of anhydrous copper sulfate was shaken for 5 minutes with 300 ml of copper-free gasoline, the copper content of 100 ml of filtered gasoline was 1 microgram. The remaining 200 ml of the suspension was then combined with 200 ml of gasoline containing 5 ppm of metal deactivator and shaken for 5 minutes. The copper content of a filtered 100-ml portion was 4 micrograms (0.052 ppm), and after standing for 18 hours, the copper content had increased to 13.6 micrograms/100 ml (0.17 ppm).

As inorganic salts settle rapidly they are ordinarily not present in the laboratory sample. It is advisable to filter all samples before extracting copper, however, as soluble copper is the principal interest, and this precaution will ensure consistent results.

The determination of copper in finished gasoline requires a somewhat more elaborate procedure than is provided here because of the large number of additives present. Common additives for gasoline include anti-detonates, chamber conditioners, scavengers, dyes, gum-inhibitors, detergents, anti-icing compounds, corrosion inhibitors, and metal deactivators. Livingstone and Lawson [13] have described a procedure in which normal amounts of iron, sulfur, and additives do not interfere. They point out that $N/10$ hydrochloric acid is more effective than $4N$ for extracting copper from gasoline; this is probably because of the greater hydrogen-ion activity in the more dilute solution.

In handling the samples under consideration it has been found that virtually complete recovery of copper is obtained by a single extraction with 25 ml of either $N/10$ or $1N$ hydrochloric acid. The acid extract is made alkaline with ammonium hydroxide, ammonium citrate is added to prevent interference by iron, a solution of sodium diethyldithiocarbamate is added, the copper salt of the latter is extracted with a 15-ml portion of carbon tetrachloride, and the transmittance of the solvent is determined at 440 mμ.

When preparing the standardization curve it is important to use the same aqueous volume as will be obtained in actual determinations, and it is especially important to use a fresh solution of sodium diethyldithiocarbamate. The gradual deterioration of this reagent causes the reagent blank to increase as time goes by, and it should be at a minimum when the calibration is made. The transmittance of a blank solution prepared with fresh reagent should be not less than 90%. It is advisable to use demineralized water for preparing all solutions of the reagent.

PREPARATION

1. REAGENTS:

   Ammonium citrate (dibasic): 20 grams in 100 ml of solution.

   Sodium diethyldithiocarbamate: Eastman No. 2596, 0.1 per cent in water. Store in a brown bottle.

   Standard copper solution: Dissolve 0.391 gram of $CuSO_4 \cdot 5H_2O$ in water, add 5 ml of $1:1H_2SO_4$, and dilute to 1000 ml with water. 1 ml $= 0.10$ mg Cu^{++}.

2. STANDARDIZATION:

Dilute the prepared standard copper solution tenfold to produce a solution containing .01 mg Cu/ml. To each of six

125-ml separatory funnels, add 25 ml of $N/10$ HCl, and transfer .01, .02, .03, .04, and .05 mg of Cu^{++} to the funnels, using the sixth as a blank. Add 2 ml of NH_4OH, 5 ml of 20% ammonium citrate, and 1.0 ml of 0.1% sodium diethyldithiocarbamate, mixing after each addition.

With a pipette transfer 15.0 ml of carbon tetrachloride to each funnel and shake it for 2 minutes. Allow the layers to separate, and filter the CCl_4 layer through a plug of glass wool in the funnel stem into a dry 13-mm cuvette. Measure the transmittances at 440 mμ without undue delay and plot the transmittances obtained against milligrams of copper on semi-logarithmic paper.

NOTE: The color fades slowly in the light.

PROCEDURE

1. Filter the sample of gasoline, and transfer a 200-ml portion to a 250-ml separatory funnel. Add 25 ml of $N/10$ HCl and shake the funnel for 2 minutes. Allow the layers to separate and transfer the acid layer to a 125-ml separatory funnel, add 2 ml of NH_4OH, 5 ml of 20% ammonium citrate, and 1.0 ml of 0.1% sodium diethyldithiocarbamate, mixing after each addition.

2. With a pipette transfer 15.0 ml of CCl_4 to the funnel and shake it for 2 minutes. Allow the layers to separate, and filter the CCl_4 layer through a plug of glass wool in the funnel stem into a dry 13-mm cuvette. Measure the transmittance at 440 mμ, and read the corresponding mg of copper from the prepared calibration curve. A blank should be run each day and deducted from the result obtained for the solution of the sample.

$$\text{ppm Cu} = \frac{\text{net mg Cu} \times 10^6}{\text{ml sample} \times \text{sp gr} \times 10^3}$$

NOTES: If metal deactivator is present, it should be extracted by HCl; and the solution is colorless. Upon addition of NH_4OH, the free base is formed, producing a yellow color. The base is not appreciably soluble in CCl_4 and only a negligible amount is transferred to the CCl_4 layer in a single extraction.

If the blank exceeds 3 micrograms, a new reagent solution should be prepared.

2. Determination of Copper in Naphthenate Driers

Copper naphthenate driers are manufactured by a process similar to that described in the chapter on cobalt. The determination of copper is required for manufacturing control, and possible contaminants include calcium, cobalt, iron, lead, manganese, and zinc. Of these, only iron oxidizes iodide, and this is prevented in the procedure given here by the addition of ammonium bifluoride. As is the case with all liquid driers, the solvent is volatile, so Lunge weighing bottles or gelatine capsules must be used for weighing samples.

Organic material may be destroyed by direct wet oxidation, or by pyrosulfate fusion, as outlined in Chapter 2, Section 4. If it is desired to use the fusion procedure, reference should be made to the chapter on zinc for details of preparing the crucible and making the fusion. It may be mentioned, however, that it is usually unnecessary to add sulfuric acid during the fusion when oxidizing copper naphthenate, as there seems to be little tendency to form sulfide. When a clear melt is obtained it is cooled, dissolved in water, acetic acid is added, and the iodometric procedure followed as described here.

When the direct wet-oxidation method is applied, the sample must be dispersed in sulfuric acid, and treated with nitric acid before heating, to prevent spattering. Wet oxidation is a little slower than pyrosulfate fusion because the excess of sulfuric acid must be evaporated after the oxidation is completed. Copper naphthenate is especially difficult to oxidize completely, and the residue is often dirty after the evaporation. As the presence of nitric acid or peroxide is not permissible in the iodometric determination, it is advisable to finish the oxidation with perchloric acid; the latter usually leaves a clean residue upon evaporation.

As noted in Chapter 2, Section 1, copper is likely to attack both platinum and porcelain, so direct ashing is not applicable. Weatherburn et al.[14] note that low results are obtained by direct ashing, and they recommend acid hydrolysis for copper and zinc naphthenates. Their procedure was subsequently modified somewhat by McLeod.[15] If extraction is preferred, reference should be made to these articles for details.

PROCEDURE

1. Transfer a suitable weighed sample to a 400-ml beaker, cover with a ribbed watch glass, add 10 ml of H_2SO_4, and 5 ml of HNO_3, and allow the initial vigorous reaction to subside. Oxidize the sample in the usual way (Chapter 2, Section 3), and when the solution is clean, add 5 ml of HNO_3 and 5 ml of $HClO_4$. Heat the solution until these acids have boiled off, and evaporate the sulfuric acid to a volume of about 2 ml. If there is any darkening of the solution, repeat these acid treatments. Finally, place an asbestos sheet under the beaker and continue the evaporation to dryness.

> NOTES: The amount of sample should be sufficient to contain 60–75 mg of copper.
> As copper naphthenate is difficult to oxidize completely more perchloric acid than usual is specified.
> The asbestos sheet moderates the heat and prevents decrepitation as the residue approaches dryness.

2. When no more fumes are evolved, cool the beaker, and dilute to about 150 ml with water. Add 10 ml of HAc, 1 g of NH_4HF_2, and mix thoroughly. Add 3 g of KI, and titrate with standard $N/10$ $Na_2S_2O_3$ until the iodine color is indistinct. Add 5 ml of starch solution and 4 g of KSCN, and complete the titration to the disappearance of the blue color. Discard the solution without delay, and calculate the percentage of copper in the sample.

$$\% \text{ Cu} = \frac{\text{ml } S_2O_3^= \times N \times 63.54 \times 100}{\text{sample wt} \times 1000}$$

> NOTES: Ammonium bifluoride is used to prevent the oxidation of iodide by small amounts of ferric ion which may be present.
> Potassium thiocyanate should be used rather than the ammonium salt, as the latter is subject to decomposition and may participate in abnormal reactions with iodine.[9] The salt should not be added until near the end point as it is oxidized by iodine.
> The hydrofluoric acid etches the beaker if allowed to stand very long.

3. Determination of Copper in Distillates

As copper has an undesirable effect on cracking cat-alysts,[16] the determination of small concentrations of the element in distillates is frequently required. Karchmer and Gunn [6] have described a spectrographic method, and Milner et al.[17] determine copper polarographically in these stocks after wet ashing.

The procedure here utilizes the method of soft ashing and wet oxidation after pretreatment with a few drops of fuming sulfuric acid. The presence of metallic elements in distillates is presumptive evidence that they are present as volatile organic compounds, and, as discussed in Chapter 2, Section 2, the addition of sulfuric acid effectively decom-poses them, and metallic elements are retained in the ash.

The small concentrations of copper in distillates, 0.05–0.5 ppm, are conveniently determined by the carbamate colori-metric procedure, and several possible interfering elements should be considered. The metal contents of about twenty heavy distillates from six different crudes were in the fol-lowing ranges: chromium, 0.01–0.05 ppm; iron, 0.2–3 ppm; lead, less than 0.1 ppm; nickel, 0.05–1 ppm; vanadium, 0.05–2 ppm. None of these interferes in the procedure when present in these concentration ranges, and no separations are required. A detailed discussion of the carbamate method is given by Sandell,[18] and reference should be made to his book for further details.

PREPARATION

1. REAGENTS:

   The same reagents are required for the following procedure as for that in Section 1.

2. STANDARDIZATION:

Dilute the prepared copper standard solution tenfold to produce a solution containing 0.01 mg Cu/ml. To each of six 125-ml separatory funnels, add about 50 ml of water con-taining 3 ml of H_2SO_4. Add 2 drops of phenolphthalein indicator, and neutralize to a permanent pink with NH_4OH, adding 2 ml excess. Transfer .01, .02, .03, .04, and .05 mg of Cu^{++} to the funnels, using the sixth as a blank. Add 5 ml of 20% ammonium citrate, and 1.0 ml of 0.1% sodium diethyldithiocarbamate, mixing after each addition.

Continue with the carbon tetrachloride extraction, etc., as in Section 1, *Standardization of Copper Solution,* and plot the transmittances obtained against milligrams of copper taken.

PROCEDURE

1. If the sample is a heavy gas oil, warm it in a hot water bath to reduce its viscosity, and mix thoroughly. Transfer a suitable sample (100–200 grams) to a clean dry 400-ml beaker. Place the beaker on a steam plate, add 10 drops of fuming sulfuric acid. Stir the sample thoroughly, and heat for about 10 minutes with occasional stirring. Remove the stirring rod, wipe it with a piece of filter paper, and drop the paper into the beaker.

2. Place the beaker in a muffle can (see Chapter 2, Section 2), and ignite the sample with a gas flame. As the sample burns down it may be necessary to apply the flame to maintain burning. Do this by placing the burner near the can, but do not heat the bottom directly. When the sample concentrates to a heavy tar, or soft ash, burn the loose carbon from the walls of the beaker with a flame as usual. Cool the beaker, add about 20 ml of H_2SO_4, and allow the mixture to digest on the hot plate for at least 15 minutes. Oxidize as usual with HNO_3, finishing with H_2O_2 and $HClO_4$, evaporating the sulfuric acid to a residual volume of 2–3 ml.

NOTE: Some samples may require longer digestion with more sulfuric acid than here specified. It is recommended that reference be made to Chapter 2, Sections 2 and 3, when using this procedure.

3. Cool the acid solution, dilute it to about 30 ml with water, and boil gently for 5 minutes. Cool, filter if necessary, and transfer it to a 125-ml separatory funnel, diluting in the process to about 50 ml. Add 2 drops of phenolphthalein, neutralize to a permanent pink with NH_4OH, and add 2 ml in excess. Cool, if necessary, add 5 ml of 20% ammonium

citrate, and 1.0 ml of 0.1% sodium diethyldithiocarbamate, mixing after each addition.

NOTE: A small amount of silica occasionally separates and must be filtered.

4. Transfer 15.0 ml of CCl_4 to the funnel and shake it for 2 minutes. Allow the layers to separate, and filter the CCl_4 layer through a plug of glass wool in the funnel stem into a dry 13-mm cuvette. Read the transmittance at 440 mμ without undue delay, and read the corresponding milligrams of copper from the prepared calibration curve.

$$ppm\ Cu = \frac{mg\ Cu \times 10^6}{sample\ wt \times 10^3}$$

NOTE: It is advisable to carry a blank through the procedure.

4. Determination of Copper in Used Lubricating Oils

Elements that are commonly found in used lubricating oils in significant amounts include aluminum, barium, calcium, chromium, copper, iron, lead, nickel, phosphorus, silicon, sulfur, and zinc. For reasons discussed in Chapter 2, the ashing procedures are not very convenient, and wet oxidation in sulfuric acid leaves barium, calcium, chromium, lead, and silicon in insoluble forms, heavily contaminated by coprecipitation and mechanical inclusion of normally soluble salts.

Treating the sample with an alcoholic solution of ammonium sulfide as outlined in Chapter 2, Section 10, converts copper, lead, and zinc to their sulfides, and these metals can be quantitatively recovered by filtration. This method, not previously described in the literature, is very convenient for recovering both copper and lead from petroleum oils.

The sample is dissolved in a mixed solvent, paper pulp is added as a collector, and the mixture is heated with alcoholic ammonium sulfide. The mixed precipitates of copper, lead, and zinc sulfides (iron sulfide is partially precipitated also) are filtered on paper and washed successively with acetone and hexane. The precipitate and paper are returned to the original beaker and dried on the steam plate. After wet oxidation, copper is determined iodometrically.

The work of Waldbauer *et al.*,[19] has shown that copper is coprecipitated with lead sulfate, probably by adsorption or occlusion. For this reason lead sulfate cannot be filtered out of the oxidized solution and discarded, nor is it possible to evaporate the sulfuric acid to dryness, as in Section 2, because lead sulfate tends to fuse into the bottom of the beaker. In the following procedure, sulfuric acid is evaporated to 2–3 ml, diluted, and sodium acetate is added. Upon addition of potassium iodide, lead sulfate is transformed to the less soluble iodide, thus releasing the occluded copper.

When handling badly oxidized oils the maximum sample that can be treated conveniently is about 20 grams; as much as 50 grams can be used if the sample is in good condition. Sufficient mixed solvent should be used to produce an easily filterable solution; three to four times the weight of the sample in milliliters is usually satisfactory.

PREPARATION

1. REAGENTS:

Mixed solvent: Mix equal volumes of benzene, acetone, and hexane (or petroleum ether).

Sulfide reagent: Mix 400 ml of Formula 30 alcohol with 100 ml of NH_4OH, and bubble a rapid stream of H_2S into the solution for 5 minutes. Store in a glass bottle with a rubber stopper.

PROCEDURE

1. Weigh a suitable sample, say 20 grams, and transfer it to a 250-ml beaker. Add 50 ml of mixed solvent and mix thoroughly. Add a small amount of paper pulp, disperse it by vigorous stirring, and while continuing to stir, add 25 ml of sulfide reagent. Heat this to boiling on the steam plate, continuing to stir. Set it aside until the precipitate settles, and filter on Whatman No. 41 paper. Wash the beaker and precipitate, first with acetone, then with hexane (or petroleum ether), and discard the filtrate.

NOTES: By adding the pulp first, the mixed sulfides are precipitated in the interstices of the paper and filtration is facilitated.

The paper may be fitted to the funnel by using a thinner to wet the paper.

2. Transfer the paper and precipitate to the original beaker and heat it on the steam plate until no odor of solvent remains. Add sulfuric acid and oxidize the paper and precipitate in the usual way, evaporating to a final volume of 2–3 ml of H_2SO_4.

3. Cool the solution, dilute to 150 ml with water, and boil for about 5 minutes. Cool, add 10 grams of NaAc, 1 gram of NH_4HF_2, and titrate with standard $N/10$ $Na_2S_2O_3$ until the iodine color is indistinct. Add 5 ml of starch solution and 4 grams of KSCN, and complete the titration to the disappearance of the blue color. Discard the solution without delay and calculate the percentage of copper in the sample, as in Section 2.

NOTE: Iron is partially precipitated as sulfide and bifluoride must be used to prevent its interference.

References

1. M. C. K. JONES and R. L. HARDY, "Petroleum Ash Components and Their Effect on Refractories," *Ind. Eng. Chem.*, 44:2615 (1952).
2. H. BUCHWALD and L. G. WOOD, "The Determination of Copper in Mineral Oils Using an Ion-Exchange Technique," *Anal. Chem.*, 25:664 (1953).
3. D. M. ZALL, R. E. McMICHAEL, and D. W. FISHER, "Determination of Copper in Fuel Oil and Other Petroleum Products," *Ibid.*, 29:88 (1957).
4. J. E. BARNEY, II, "Spectrochemical Determination of Copper in Turbine Oils," *Ibid.*, 26:567 (1954).
5. C. R. HODGKINS and J. HANSEN, "Spectrochemical Determination of Copper in Crankcase Drainings," *Ibid.*, 26:1759 (1954).
6. J. H. KARCHMER and E. L. GUNN, "Determination of Trace Metals in Petroleum Fractions," *Ibid.*, 24:1733 (1952).
7. H. W. FOOTE and J. E. VANCE, "A Comparison of Quantitative Methods for the Determination of Copper," *Ind. Eng. Chem., Anal. Ed.*, 9:205 (1937).
8. W. R. CROWELL, "Iodometric Determination of Copper," *Ibid.*, 4:159 (1939).
9. E. W. HAMMOCK and E. H. SWIFT, "Iodometric Determination of Copper," *Anal. Chem.*, 21:975 (1949).
10. I. M. KOLTHOFF, R. BELCHER, V. A. STENGER, and G. MATSUYAMA, *Volumetric Analysis*, Volume III, p. 347, Interscience (1957).
11. L. SILVERMAN, "Sulfamic Acid as an Aid in the Analytical Elec-

trodeposition of Copper," *Ind. Eng. Chem., Anal. Ed.*, 17:270 (1945).

12. O. T. ZIMMERMAN and I. LAVINE, *Handbook of Material Trade Names*, Supplement II (1957), Industrial Research Service, Inc., Dover, New Hampshire.

13. J. K. LIVINGSTONE and N. D. LAWSON, "Photometric Determination of Copper in Gasoline," *Anal. Chem.*, 25:1917 (1953).

14. A. S. WEATHERBURN, M. W. WEATHERBURN, and C. H. BAYLEY, "Determination of Copper and Zinc in Their Naphthenates," *Ind. Eng. Chem., Anal. Ed.*, 16:703 (1944).

15. A. A. McLEOD, "Determination of Copper in Copper Naphthenate," *Ibid.*, 17:599 (1945).

16. G. A. MILLS, "Aging of Cracking Catalysts," *Ibid.*, 42:182 (1950).

17. O. I. MILNER, J. R. GLASS, J. P. KIRCHNER, and A. N. YURICK, "Determination of Trace Metals in Crudes and Other Petroleum Oils," *Anal. Chem.*, 24:1728 (1952).

18. E. B. SANDELL, *Colorimetric Determination of Traces of Metals*, 2nd Ed., Interscience, New York (1950).

19. L. WALDBAUER, F. W. ROLF, and H. A. FREDIANI, "Spectrographic Studies of Coprecipitation," *Ind. Eng. Chem., Anal. Ed.*, 13:888 (1941).

Chapter 11

THE HALOGENS

With the exception of iodine, the halogens are of considerable importance in petroleum refining, and many methods have been proposed for determining them in a variety of products. As iodine is of little significance it will not be considered. Bromine is frequently determined in gasoline, but is ordinarily not found in other products. The determination of fluorine is required for the operation of hydrofluoric acid alkylation plants, but the element has no application as a petroleum additive. By far the most important halogen is chlorine; it is present in many additives, and it is also an undesirable contaminant in some phases of refining.

Chlorine is used as an additive in a wide range of concentrations, frequently associated with lead, phosphorus, selenium, and sulfur. Products that may contain chlorine additives include lubricating oils, gear lubricants, various greases, and lubricants for high load bearings, as well as specialty products such as cutting oils and wood preservatives. Ethylene chloride is, of course, a familiar gasoline additive.

The presence of chlorine is undesirable in some products, such as insulating oils, cable oils, xylenes, and phenols; and in feed stocks for certain catalytic reactions whose kinetics are affected by its presence. In some instances even very small amounts are objectionable.

One of the most widely-used procedures for determining chlorine in petroleum products is the oxygen bomb method.[1] As this is an established routine procedure, it will not be considered in detail in this chapter. The method is applicable to new and used lubricating oils and greases, and to most chlorine additives. The method is infrequently applied to greases, however, as these products are usually qualified by physical tests, such as penetration, dropping point, bearing throwout tests, and the like; greases are seldom subjected to chemical analysis.

As a number of different products and halogen concentrations are to be covered in this chapter, it will be divided into sections as follows:

1. Peroxide Bomb Combustion
2. Sodium Dehalogenation
3. Extraction Procedures
    a) Reaction with Sodium Biphenyl
    b) Extraction of Inorganic Chlorides
    c) Determination of Salts in Crude Oils
4. Wickbold Oxyhydrogen Flame Combustion
    a) Decomposition of Organic Fluorine Compounds
    b) Decomposition of Organic Chlorine Compounds
5. Potentiometric Determination of Bromide and Chloride
6. Colorimetric Determination of Chloride
7. Volumetric Determination of Fluoride

Table 11:1 summarizes typical products, approximate halogen contents, and appropriate methods of decomposition and determination; the numbers refer to sections of this chapter.

TABLE 11:1

HALOGENS IN PETROLEUM PRODUCTS

Product or Stock	Halogen Concentration	Method of Decomposition	Method of Determination
Crude Oils, Chloride	0–150 ppm	3(c)	5
Gasolines, Bromine	250–800 ppm	3(a)	5
Chlorine	0–400 ppm	3(a)	5
HF-Alkylation Stocks, Fluorine	0–500 ppm	4(b)	7
Insulating and Cable Oils, Chlorine	0–100 ppm	2,4(b)	5,6
Miscellaneous Lubricants, Chlorine	0.3–3%	(*),2	5

TABLE 11:1 (Cont.)

HALOGENS IN PETROLEUM PRODUCTS

Product or Stock	Halogen Concentration	Method of Decomposition	Method of Determination
Naphtha Feed Stocks, Chlorine	0–10 ppm	2,4(b)	6
Selenium- Chlorine Additives	0.1–1.5%	1,*3(a)*	5
Sulfur-Chlorine Additives	3–45%	1,*3(a)*,4(b)	5
Sulfur-Lead- Chlorine Additives	3–5%	1,*3(a)*	5
Sulfur- Phosphorus- Chlorine Additives	25–30%	1,*3(a)*	5
Wood Preserva- tives, Chlorine	0.5–1%	(*)	5

* These products are best handled by the oxygen bomb method. Italics indicate preferable method.

1. Peroxide Bomb Combustion

The peroxide bomb fusion method has been discussed briefly in Chapter 2, Section 6. In it a small sample is combined with a prepared fusion mixture, sealed in a special bomb, and ignited electrically. Fluorine and chlorine are converted to halides; bromine and iodine are oxidized to bromate and iodate, respectively. Although carbon-fluorine bonds in aromatics—and in aliphatics with more than one fluorine attached to a carbon atom—are extremely stable, they are completely decomposed in a peroxide fusion. The method is not especially useful for the determination of fluorine, however, because of several disadvantages, including limited size of sample, introduction of large amounts of salts, and the volatility of many fluorine compounds.

It will be noted in Table 11:1 that the peroxide bomb may be used for the determination of chlorine in additives that also contain lead, phosphorus, selenium, and sulfur in various combinations. None of these interferes in the procedure given here, nor in the potentiometric titration in Section 5.

As the maximum sample is only about 0.3 gram, it will be noted that titrations are rather small for chlorine contents less than ten per cent, and it is therefore preferable to use the sodium biphenyl method in Section 3(a) for such materials. Even under the best conditions the precision of the peroxide bomb method leaves something to be desired.

It was mentioned in Chapter 2, Section 6, that gelatine capsules may be used for weighing samples, and that they should be crushed after being placed in the fusion mixture. When handling chlorine additives, however, it is permissible to center the capsule in the fusion mixture in the bomb without crushing, as some of these materials are acidic, and ignite the fusion mixture on contact.

The maximum combustible material that can be ignited safely is 0.5 gram, and as it is usually necessary to use a combustion aid, such as sucrose or benzoic acid to obtain complete decomposition, the maximum amount of sample is limited to about 0.3 gram. A small amount of potassium nitrate is used as an accelerator for samples difficult to decompose. As the method is inherently hazardous, the following safety precautions should be observed.

SAFETY PRECAUTIONS

1. The eyes, face, and hands must be protected while preparing the fusion mixture. The mixture containing potassium nitrate is explosive; not more than 0.5 gram of the salt should be used.
2. The full 15 grams of sodium peroxide must always be used.
3. The sample must be dry, and non-acidic, or else enclosed in a gelatine capsule.
4. Not more than 0.5 gram of total combustible material should be used.
5. Fusion cups should be discarded when the walls become eroded.
6. The gasket must be in good condition, and produce a tight seal between the fusion cup and the upper edge of the bell body. Most explosions are caused by faulty gaskets.
7. The ignition switch should be located several feet from the bomb.

PREPARATION

1. APPARATUS:

Parr Peroxide Bomb, and accessories, 22-ml electric-ignition type

Ignition Transformer, with ignition switch
Fuse Wire
Water Bath, circulating

2. FUSION MIXTURE FOR HALOGENS:

Prepare this mixture, *immediately before using,* by combining in a small glass-stoppered bottle, 15 grams of Na_2O_2, 0.5 gram of powdered KNO_3, and 0.2 gram of powdered sugar, or benzoic acid. Mix rapidly and thoroughly with a stirring rod, and stopper to prevent absorption of moisture. Use as soon as possible.

PROCEDURE

1. Attach a 7-cm length of fuse wire to the terminals on the bomb cover so that the wire loop will extend a short distance into the fusion mixture when the cover is in place.

2. Weigh a maximum of 0.3 gram of the sample, and transfer it to a gelatine capsule (size 00). *Using tongs or forceps—not the fingers!*—transfer approximately half of the prepared fusion mixture to the dry fusion cup, and place the closed capsule upright in the fusion mixture. Add the balance of the fusion mixture, and tap the cup lightly to pack the charge, then dust a *small amount* of sugar, or benzoic acid on the surface.

> NOTES: Be sure that there is no peroxide clinging to the walls near the top of the cup; if any is in contact with the gasket, the bomb may rupture when ignited.
>
> A very small amount of combustion aid on the surface of the charge will ensure ignition.

3. Place the cover in position, adjusting the fuse wire so that it dips a short distance into the fusion mixture. See that the cover is properly seated, and that the gasket seals both the cup and the upper edge of the bell body. Tighten the screw cap firmly with a wrench, and place the bomb in the water bath so that the ignition arm makes contact with the bomb head assembly. Ignite the charge with the ignition transformer, and allow the bomb to remain in the bath for about 10 minutes, or until cool. Remove the assembly from the bath, and wipe off the excess water.

4. Open the bomb, and if the fusion appears satisfactory,

place the cup on its side in a 400-ml beaker, add 100 ml of water, and immediately place a cover glass on the beaker. When the initial vigorous reaction subsides, heat gently until the melt dissolves, then remove the cup and wash it thoroughly with water. Dilute the solution to a volume of 200 ml, and boil this for about 15 minutes, or until the excess peroxide has decomposed.

NOTE: A few particles of carbon may be disregarded, but if any unburned material is visible the ignition must be repeated with another sample.

5. If bromine (or iodine) is to be determined, add 1 gram of hydrazine sulfate, and boil gently for about 15 minutes to reduce bromate (or iodate). Cool the alkaline solution in an ice bath, and slowly neutralize to methyl orange with 1:1 HNO_3, finally adding 5 ml excess.

6. Determine the halide potentiometrically using the procedure in Section 5.

NOTES: The addition of hydrazine is unnecessary when only chlorine is present.

The initial potential, and magnitude of the break are as given in Table 11:2 (see Section 5), but the range of the break is lower because of the large amount of salt in this solution. Burette readings should be taken starting at about 400 mv for chloride.

2. Sodium Dehalogenation

This method of sodium dehalogenation, for converting organically combined halogen to ionic form, has been described briefly in Chapter 2, Section 7, and several references are given there dealing with its origin and subsequent modifications. The procedure may be applied to insulating and cable oils, lubricating oils, certain greases, and various naphtha feed stocks. It is unsatisfactory for materials that consume excessive amounts of sodium, and low results may be obtained with volatile materials. In addition, lubricants that contain fats, and some heavy metals, may yield variable results. As most of the samples for which this method is used contain only chlorine, the discussion will be confined to this element.

Sodium dehalogenation is especially convenient for recovering organic chlorine from polymeric material, such as cable and insulating oils, and from naphtha feed stocks for the re-forming by catalysts. Pennington *et al.*[2] have described an amperometric determination of micro amounts of chlorine in petroleum naphtha after sodium dehalogenation. If amperometric equipment is not available these stocks can be burned in the Wickbold apparatus (see Section 4(b)), and the chloride determined colorimetrically by the procedure in Section 6. The potentiometric titration in Section 5 cannot be used for naphthas that contain less than about 10 ppm chlorine, as the practical limit for a sample is about 100 grams.

As remarked before, chlorine is undesirable in the feed stocks for several catalytic reactions, and it is also a factor in the corrosion of reactor vessels. Some cracked naphthas contain organic sulfur which reacts with sodium to yield sulfide. Hurd and Wilkinson[3] have shown that organic chlorosulfides yield mercaptans (thiols). Both sulfide and mercaptides interfere in the potentiometric titration with silver nitrate, but they are easily oxidized and rendered innocuous. Mercaptans are oxidized by peroxide to disulfides; and nitric acid oxidizes both mercaptans and disulfides to the corresponding alkylsulfonic acids. Sulfide is rapidly oxidized to sulfur by either peroxide or nitric acid.

$$2\ RSH + H_2O_2 = RSSR + H_2O$$
$$3\ RSSR + 10\ HNO_3 = 6\ RSO_2OH + 10\ NO + 2\ H_2O$$

After reaction of the sample with sodium, the liberated chloride, sulfide, and mercaptides are extracted with water, the combined extracts are acidified with nitric acid, peroxide is added, and the solution is boiled to oxidize the sulfur compounds. The extraction is not essential when determining chlorine in cable oils, but these products are extremely viscous, their stringy consistency results in coated electrodes, and vigorous stirring is required to attain equilibrium at the silver electrode. It is therefore convenient to separate the organic phase before titrating; a few milliliters of ethyl ether will prevent emulsification.

PREPARATION

1. APPARATUS:

Flask: 300-ml Erlenmeyer, with standard-taper glass joint.

Reflux condenser: 24-inch water-jacketed, with standard-taper glass joint.

PROCEDURE

1. Weigh and transfer a suitable sample to a clean dry 300-ml Erlenmeyer flask. If the sample is viscous add 50 ml of benzene. Add 5 ml of isopropyl alcohol, and two grams of metallic sodium, cut into $\frac{1}{8}$-inch cubes.

> NOTES: For naphthas, and similar products, a 100-gram sample is satisfactory; for cable oils it may be convenient to use a factor weight (35.5 grams).
>
> The addition of benzene is not necessary for naphthas, or other solvent-type materials.

2. Connect the flask to the reflux condenser, and heat on a hot plate, allowing the mixture to reflux for 30 minutes. Add through the top of the condenser a second 5-ml portion of isopropyl alcohol, and continue to reflux for an additional 30-minute period. Destroy the excess of sodium by slowly adding 25 ml of methanol through the condenser, and continue boiling until the metal has disappeared.

> NOTES: The second portion of alcohol is added to clean the surface of the sodium.
>
> Sodium reacts much more smoothly with methanol than with water.

3. After cooling, disconnect the flask and transfer the mixture to a separatory funnel, rinsing the flask with about 50 ml of water. Add 5–10 ml of ethyl ether, shake the funnel for 1 minute, and draw off the aqueous layer into a 400-ml beaker. Repeat the extraction with two additional 15-ml portions of water. Acidify the combined extracts to methyl orange with 1:1 HNO_3, add 2–3 ml of 30% H_2O_2, and heat on the steam plate until the ether has evaporated, and then boil for 5 minutes.

NOTES: Polymeric materials tend to form emulsions, but the addition of ether is unnecessary with naphthas, and the like.

This oxidation step is not required when determining chlorine in insulating and cable oils.

4. Cool, dilute to about 200 ml, add 5 ml of 1:1 HNO₃, and titrate the solution as described in the procedure in Section 5.

NOTE: If it is likely that less than 2 mg of chloride is present, the solution should be titrated with 0.01 N AgNO₃, added in 0.5-ml increments.

3. Extraction Procedures

In this section some methods that utilize separatory funnels are described for extracting halogens from petroleum. The most important of these, the reaction with sodium biphenyl, is considered first. Next the problem of extracting inorganic chlorides from petroleum is considered in a general way, and finally the extraction of chlorides from crude oils is covered in detail.

A. REACTION WITH SODIUM BIPHENYL

The reaction of sodium biphenyl with organic halides was proposed by Pecherer et al.,[4] for determining bromine and chlorine in gasoline. Ethylene bromide and ethylene chloride are added to tetraethyllead fluids as lead scavengers to prevent engine deposits. The preparation of the reagent was modified by Liggett,[5] with an improvement in stability.

The reaction is essentially instantaneous, removing halogen completely from such stable compounds as carbon tetrachloride, hexabromobenzene, and hexachlorobenzene. It is necessary to prepare the reagent in "activating" solvents such as ethylene glycol dimethyl ether, or tetraethylene glycol dimethyl ether; the former yields a more stable reagent. The solvents should be fairly dry (less than 0.3% water); drying with barium, or calcium oxide overnight is a satisfactory procedure. The concentration of active sodium biphenyl, prepared as described here, is about 1 N, but the reagent is still useful at a concentration as low as 0.25 N.

The strength decreases gradually with age, but the reagent is quite stable if stored under refrigeration.

Reference to Table 11:1 shows that this procedure is suited for the determination of bromine and chlorine in gasolines, and for chlorine in additives that contain lead, phosphorus, selenium, or sulfur. Once a supply of reagent has been prepared, the procedure is safer, more precise, and more convenient than the peroxide bomb.

The weighed sample is transferred to a dry separatory funnel, the prepared reagent is added, and the mixture shaken briefly. Excess of reagent is destroyed by addition of water, and the organic phase is extracted with dilute nitric acid. The volume of the combined aqueous extracts is reduced to about 25 ml, the solution is acidified, and filtered to remove any insoluble tars, or other foreign material. Sodium peroxide is next added to destroy any residual organic material in solution. Chloride is not oxidized, but, as indicated by the following potentials, bromide and iodide are oxidized to bromate and iodate by peroxide in alkaline solution.

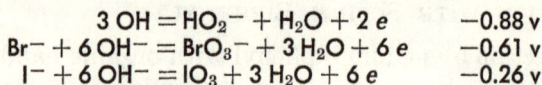

$$3\,OH = HO_2^- + H_2O + 2\,e \qquad -0.88\,v$$
$$Br^- + 6\,OH^- = BrO_3^- + 3\,H_2O + 6\,e \qquad -0.61\,v$$
$$I^- + 6\,OH^- = IO_3 + 3\,H_2O + 6\,e \qquad -0.26\,v$$

The solution is treated with a solution of sodium arsenite to reduce bromate (or iodate) and boiled. After dilution and acidification the halides are titrated potentiometrically by the procedure in Section 5.

PREPARATION

    1. REAGENTS:

        Biphenyl: Eastman No. 721.

        Ethylene glycol dimethyl ether (Ansul Chemical Company, Marinette, Wisconsin): Dry over lime or barium oxide, and filter before use.

        Nitric Acid: Dilute 20 ml of concentrated nitric acid to 100 ml with water.

        Sodium arsenite: 10% solution. Dissolve 6 grams of NaOH in water, add 6 grams of As_2O_3, stir until dissolved, and dilute to 100 ml with water. Prepare a fresh solution every two weeks.

Sodium metal: Reagent grade.

Toluene: Dried over lime or barium oxide, and filtered.

2. PREPARATION OF DISODIUM BIPHENYL REAGENT:

Transfer 300 ml of dry toluene and 58 grams of metallic sodium to a 2-liter, 3-necked flask, equipped with a heating mantle, nitrogen gas inlet, mercury seal stirrer, and reflux condenser. Heat until the toluene refluxes and the sodium melts completely. Start the stirrer, and stir until the sodium is finely dispersed. Cool to less than 10°C in a suitable bath *(not water)*. Remove the condenser, and add 1250 ml of dry ethylene glycol dimethyl ether. While stirring and passing nitrogen gas over the mixture, add 390 grams of biphenyl. The reaction starts immediately, as evidenced by the green color of sodium biphenyl; the temperature of the reacting mixture should be kept below 30°C. When the reaction is complete (1½–2 hours), pour the reagent into dry 500-ml brown prescription bottles, with screw caps, and foil liners. The reagent is stable for several months if refrigerated.

NOTE: If any unreacted sodium remains in the reaction flask, add 100 ml of isopropyl alcohol, and place the flask in a hood until the metal has dissolved.

PROCEDURE

1. Transfer 50 ml of benzene to a dry 250-ml separatory funnel, and add approximately 25 ml of the prepared sodium biphenyl reagent. If the sample is gasoline, measure its temperature, and transfer a 100-ml portion to the funnel with a pipette. Lubricating oil additives, and other samples, are weighed, and dissolved in a suitable solvent before transferring to the funnel. Stopper the funnel, and shake vigorously for 30 seconds.

NOTES: When determining chlorine in additives containing phosphorus, selenium, or sulfur, account must be taken of the fact that these elements also consume the reagent. If too large a sample is taken there may be insufficient reagent for complete recovery of halogens.

The mixture in the funnel should remain green after the initial shaking. If the mixture clears, or turns gray, or brown, additional reagent should be added until a permanent green color persists.

2. Allow the mixture to stand for 5 minutes, and add 25 ml of water to decompose the excess reagent, shake briefly, and vent the funnel. Allow the layers to separate, and drain the water layer into a 400-ml beaker. Rinse the inside of the funnel with a few milliliters of water, and without shaking, drain the water into the beaker and rinse the stem. Add 20 ml of dilute nitric acid to the funnel, shake for 1 minute, and drain the aqueous layer into the beaker. Finally, extract with two 20-ml portions of water, and discard the contents of the funnel. Add a few drops of methyl orange indicator, and if the solution is not alkaline, add sodium hydroxide pellets until the indicator is yellow. Evaporate the solution to about 25 ml, cool, and acidify with nitric acid. Add a little paper pulp, and filter the solution through Whatman No. 41 paper, collecting the filtrate in a 400-ml beaker.

NOTE: The use of toluene in the reagent facilitates the separation of the aqueous layer. Additives containing sulfur and lead tend to emulsify. The addition of 2–3 ml of ethyl ether usually resolves the emulsion.

3. To the filtrate add about 5 grams of Na_2O_2, and boil it for 10 minutes to decompose the excess. Add 10 ml of sodium arsenite solution, boil briefly, and cool to room temperature. Dilute to approximately 200 ml, acidify to methyl orange with 1:1 HNO_3, and add 10 ml excess. Titrate the halides with standard silver nitrate solution by the procedure outlined in Section 5.

NOTE: Halogens in gasoline are usually expressed in grams per gallon at 60°F; these calculations are given in Section 5.

B. Extraction of Inorganic Chlorides

The determination of inorganic chlorides in petroleum is occasionally required, and without giving a detailed procedure, a few pertinent points will be considered here. The petroleum sample is diluted with a suitable solvent, if necessary, and extracted in a separatory funnel with water slightly acidified with nitric acid. As usual in this process the avoidance, or resolution of emulsions is the principal problem. The choice of solvent, and the nature of the sample are significant factors in the tendency to form emulsions, and separations are sometimes improved by using hot water, or demulsifiers, or both.

Salts may be present as a result of neutralizations in manufacturing processes, or by contamination. Soaps, such as sulfonates, are best dissolved in ethyl ether, and extracted with dilute nitric acid (1:9). The total volume of ether should be 100 ml, and three 25-ml portions of the dilute acid are sufficient to recover the chlorides. After diluting the combined extracts, the chloride is titrated by the procedure in Section 5.

Certain additives may be contaminated by chloride during manufacture, and some of these form very stable emulsions when shaken with water. This reaction can be avoided by dissolving a small sample (3–5 g) in 100 ml of benzene, heating it on a steam plate to hasten solution, and adding 50 ml of methyl alcohol to the hot solution. A quantity of distilled water, acidified to methyl orange with nitric acid, should be heated to about 140°F. The sample solution is then transferred to a separatory funnel, using about 50 ml of the hot water to make the transfer. If the sample is alkaline, 1:1 nitric acid is added by drops until the indicator is red. The funnel is then agitated gently, care being taken to release the pressure frequently. After three successive extractions with hot acidified water the combined extracts are titrated potentiometrically. The use of hot water, and the addition of methanol, are effective in preventing emulsification.

Lubricating oils can be thinned with naphtha, and extracted with acidified water; a few milliliters of normal butyl alcohol may be added to resolve emulsions. Heavier products including asphalts, fuel oils, and crudes should be handled by the procedure in the next section.

C. Determination of Salts in Crude Oils

The presence of inorganic salts in crude oils is a serious problem in refining, as their presence causes plugging of furnace tubes and deposits in condensers, lines, and auxiliary equipment, as well as high ash contents of residua. The presence of salts usually indicates that the oil contains emulsified brine, although in some instances they are also introduced from drilling muds.

Several methods have been proposed for the use of separatory funnels in extracting salts from crude oils. Blair [6] recommended xylene as a solvent with the addition of Tret-o-Lite * destabilizers to flocculate emulsified droplets, stabilized by natural emulsifying agents in the oils. Neilson, et al.[7] used xylene, benzene, or toluene as a solvent, with phenol added to liberate protected crystals. In both of these methods hot water is used for extraction.

Shaking with water alone is not sufficient to remove all salts. The use of phenol or destabilizers promotes the extraction of wax- and asphalt-coated crystals, and the coalescence of highly dispersed droplets of brine. Xylene is a desirable solvent, as it has a low vapor pressure and specific gravity, both of which properties facilitate the separation of the aqueous layer.

Extremely viscous samples (heavy fuels, asphalts, etc.) should be handled as follows: Warm the container, and pour as much of the sample as possible into a beaker. Dissolve the material in xylene, warming if necessary. Transfer the xylene solution to the separatory funnel, and wash the sample container with xylene followed by hot water, as described in the procedure.

Determinations in the extract may include calcium, magnesium, sodium, sulfate, chloride, and total salts. As the most offensive is chloride, because of the hydrolysis on heating of calcium and magnesium chlorides to produce hydrochloric acid, the procedure given here will cover only this element.

Before proceeding to the details of the extraction, attention is called to a series of articles by Stoffer,[8] describing the use of the flame photometer for estimating the salt contents of crudes. The salt contents in pounds per thousand barrels are plotted against luminosities obtained by aspirating the oil sample, diluted with an equal volume of benzene,

* Manufactured by the Tretolite Company (Division of the Petrolite Corp.), 369 Marshall Avenue, St. Louis 19, Missouri.

through the flame photometer. This is a very useful procedure for handling segregated crudes, but is not applicable to mixed crudes, or to unknown samples.

PREPARATION

1. REAGENTS:
   Phenol-benzene solution: Equal parts by weight.
   Xylene.

PROCEDURE

1. Weigh the bottle containing the sample, and warm it to about 125°F in a water bath if the sample is viscous.

> NOTES: Samples for inorganic chlorides should be taken in 4-oz oil sample bottles. Because of the difficulty of obtaining a uniform sample, it is advisable to use all of the submitted sample.
> The handling of extremely viscous samples is outlined in the discussion in Section 3(c).

2. Pour 100 ml of xylene into a 500-ml separatory funnel, and pour the sample directly into the funnel. Shake another 25-ml portion of xylene in the sample bottle, adding it to the funnel. Repeat if necessary. The total volume of solvent-plus-sample should be 250–300 ml. Dry and reweigh the sample bottle.

3. Pour 100 ml of hot water into the sample bottle. Shake the bottle and pour the contents into the separatory funnel. Add 10 ml of the phenol-benzene solution, and shake the funnel for 5 minutes, releasing the pressure occasionally. Add a 5-ml portion of the phenol solution, stir gently, and allow the layers to separate. Drain the aqueous layer into a 400-ml beaker, and extract twice more with 100-ml portions of hot water.

> NOTES: Beware of the possibility of the bottle breaking when the hot water is added.
> The phenol is added first as a solvent for asphalt, gum, or wax that encloses salt crystals; the second addition aids in resolving emulsions.
> A heating lamp may be used to aid the separation of the aqueous layers; be sure to remove the stopper from the separatory funnel.

4. Acidify the combined extracts with 1:1 HNO_3 to methyl orange indicator. Add 2ml of 30% H_2O_2, and boil 5

minutes. Cool, add 5 ml of 1:1 HNO$_3$, and titrate the solution by the procedure in Section 5.

> NOTES: Peroxide is used to oxidize any hydrogen sulfide or thiols that may be present.
>
> Salt contents are customarily expressed as pounds of NaCl per thousand barrels, and to do this the specific gravity of the sample is required. The calculation is as follows:

$$\text{NaCl, lb/1000 bbl} = \frac{\text{ml Ag}^+ \times N \times 58.45 \times 350 \times \text{sp gr}}{\text{weight of sample}}$$

4. Wickbold Oxyhydrogen Flame Combustion

The use of the Wickbold combustion apparatus has been mentioned in Chapter 2, Section 9(b) and it is described in detail in the Appendix. This section will cover the application of this method of combustion for the determination of fluorine in streams in the hydrofluoric acid alkylation process, and for chlorine in naphtha feed stocks.

A. Decomposition of Organic Fluorine Compounds

The determination of fluorine in the petroleum industry is practically confined to streams, and products from hydrofluoric acid alkylation plants. The details of this process have been described by Frey,[9] who also gives a flow diagram of a typical plant. The method consists of mixing anhydrous hydrofluoric acid with olefins, such as butylenes, propylenes, or amylenes, to form intermediate fluorine addition compounds. These are reacted with a paraffin, such as isobutane, to produce isoparaffin blending agents with concomitant splitting off of hydrogen fluoride. The recovered acid gradually becomes contaminated with dissolved high molecular weight organic material, and as the amount increases the catalytic activity is reduced, and under-reacting occurs. This is evidenced by increasing amounts of organic fluorides in the C$_3$–C$_4$ and alkylate fractions. The propane and n-butane are used in other refinery operations, and the presence of organic fluorides is undesirable as they are subsequently hydrolyzed by heat and traces of water which release hydrofluoric acid and cause corrosion. Consequently these streams are treated with bauxite to remove the organic fluorides. Any small amount (1–10 ppm) that may still remain in the alkylate is removed by a second bauxite treatment.

The Wickbold combustion method has been used for the decomposition of organic fluorine compounds by Wickbold,[10] and by Sweetser.[11] They have found that even such stable materials as Teflon, fluorobenzene, and tetrafluoromethane are decomposed completely in the oxyhydrogen flame. The organic fluorides in hydrofluoric acid alkylation streams are not particularly stable, and as these samples are light hydrocarbons, there is no difficulty in burning them in the suction burner.

It is necessary to use a cooler in drawing these samples, and the sample bottles must be kept cold with dry ice while being transported to the laboratory. As most of these alkylation plant streams are very light, the loss by weathering radically alters the sample if it is not chilled.

After the sample is decomposed and the evolved hydrogen fluoride is absorbed in dilute alkali, the fluoride is determined volumetrically by titrating with standard thorium nitrate solution. Monnier et al.[12] have described a colorimetric procedure based upon the bleaching effect of fluoride on the peroxy ion $TiO_2(SO_4)_2^{--}$.

PROCEDURE

1. Rapidly weigh a suitable chilled sample, and transfer it to a 60-ml conical weighing bottle with an outside ground cap. Charge the Wickbold absorber with 50 ml of dilute sodium hydroxide solution (4 grams/liter), and burn the sample in the apparatus as described in Chapter 2, Section 9(b), and the Appendix.

NOTE: For samples after bauxite treatment 20–30 grams should be used; for other samples 10–20 grams is usually satisfactory. An amount should be taken which will give a titration of 1–5 milliliters.

2. Drain the contents of the absorber into a Coors No. 3A porcelain casserole, and rinse the spray trap and absorber with water, adding these rinsings to the casserole. Adjust the total volume to about 175 ml, and determine the fluoride

by titrating with standard thorium nitrate using the procedure in Section 7 of this Chapter.

NOTE: No. 3A casseroles have a capacity of about 210 ml. The use of these vessels aids in judging the end point.

B. DECOMPOSITION OF ORGANIC CHLORINE COMPOUNDS

Chlorine is objectionable in certain catalytic processes, such as reforming operations that involve platinum catalysts. Halogen adsorbed on the surface of the catalyst promotes cracking and coking reactions, seriously affecting the product, and the surface of the catalyst. There is evidence that chlorine compounds used in conditioning wells are introduced into the naphtha fractions used as feed stock for catalytic reforming. The determination of chlorine in these feed stocks is therefore required to assure their suitability.

As the concentration of chlorine in naphtha is small it is necessary to decompose a large sample. The Wickbold apparatus is especially suited for this application. Wickbold [13] has described a nephelometric determination of chloride following the decomposition of the sample and absorption of hydrogen chloride. In Section 6 a colorimetric procedure using mercuric chloranilate is described, which is suitable for these naphtha feed stocks. The method may also be used for determining chlorine in benzene, toluene, xylene, and the like. When handling samples containing larger amounts of chlorine the potentiometric procedure in Section 5 should be used for the final determination.

Although it has been reported that the hydrogen chloride evolved in a Wickbold combustion is absorbed completely in water,[13] it is necessary to evaporate the solution to a volume of about 25 ml for the colorimetric determination in Section 6. If water alone is used low and erratic results are obtained by this procedure. For this reason it is advisable to add a small amount of dilute sodium hydroxide to the absorber water if the procedure in Section 6 is to be used.

PROCEDURE

1. Weigh out a suitable sample, and transfer it to an Erlenmeyer flask. Charge the Wickbold absorber with 50 ml

of distilled water and 1.0 ml of $N/10$ NaOH, and burn the sample in the apparatus as described in Chapter 2, Section 9(b), and in the Appendix.

NOTES: If the sample contains less than 10 ppm chlorine, a 100-gram sample should be burned.

Samples containing less than 100 ppm chlorine can be absorbed in water. Samples more concentrated than this are likely to yield small amounts of free chlorine, and it is advisable to add a few drops of 30% H_2O_2 to the water in the absorber. For relatively high concentrations (2%, or more) the results may be erratic, unless the sample is diluted with an appropriate solvent before burning.

2. Drain the contents of the absorber into a beaker, and rinse the spray trap and absorber with water. If the amount of chloride to be determined is small treat the solution by the procedure in Section 6. For larger amounts, dilute the solution to about 200 ml, acidify with 1:1 HNO_3, and titrate with standard silver nitrate by the procedure in Section 5.

5. Potentiometric Determination of Bromide and Chloride

After bromine and chlorine have been converted to ionic form by any of the procedures covered in Sections 1 to 4, they can usually be determined most conveniently by potentiometric titration with standard silver nitrate. A device for measuring small potential changes without draining appreciable current is required. The Beckman Model G Laboratory pH Meter * is widely used, and the availability of such an instrument is assumed in the following discussion and procedure.

For the potentiometric determination of bromide and chloride, silver is used as the indicating electrode with glass as the reference electrode. Other combinations can be used, such as silver-silver chloride, or silver-calomel (with a sodium sulfate salt bridge), but the combination recommended here is considered most satisfactory and reliable, particularly when bromide and chloride are titrated in the same solution.

Commercial glass electrodes are equipped with a shielded

* Manufactured by Beckman Instruments, Inc., Scientific Instrument Division, 2500 Fullerton Road, Fullerton, California.

terminal connector which is secured in the instrument by a set screw; metallic electrodes are connected with a pin jack. When using an instrument with both "pH" and "millivolt" scales it is advantageous to use the pH scale, because the potential "breaks" are larger; for convenience, the readings from this scale will be referred to here as "millivolts." Lykken and Tuemmler [14] have discussed the use of glass as a reference electrode in potentiometric titrations.

The silver electrode should be cleaned with emery paper and immersed briefly in 1:1 nitric acid to produce a bright surface that comes to equilibrium rapidly with the halide solution. If an insensitive electrode is used a large change in potential with the addition of the first drop of silver nitrate solution may occur.

Titrations of solutions less than $0.001N$ in chloride are uncertain because of the appreciable solubility of silver chloride at this dilution. For practical purposes, however, it is possible to obtain a reasonable estimate with as little as 3 mg of chloride present in a volume of 200 ml, although it is usually possible to prepare a large enough sample to ensure that sufficient halide will be obtained for a precise determination. For most purposes a $0.1N$ standard silver nitrate solution is satisfactory; solutions less than $0.01N$ are not practical because the potential changes become too small.

The most frequent determination of bromine is in leaded gasolines; these may or may not also contain chlorine. The pretreatment of gasolines for the determination of halogens is covered in Section 3a), and the final determinations are made by the procedure given here. In Table 11:2 the behavior of the two elements is summarized with respect to the observed potentials, and their changes during potentiometric titration with silver nitrate; these are pH scale readings using the silver-glass electrode system. They are approximate values, and are altered by acidity, foreign salts, temperature, dilution, and other factors.

The column headed *Initial Readings* gives the potentials at which it is advisable to begin adding small uniform increments of standard solution and recording the corresponding equilibrium meter readings; the other columns are self-explanatory. Note that when both bromide and chloride are titrated in the same solution the bromide break is much decreased, whereas the chloride break is a little larger than when the halides are titrated individually; the breaks also occur about 100 millivolts lower.

TABLE 11:2

POTENTIOMETRIC TITRATION OF BROMIDE AND CHLORIDE

Halide	Initial Potential	Initial Readings	Range of Break	Magnitude of Break
Bromide	125–300	350	375–700	90–200
Chloride	425–525	550	600–700	25–50
{Bromide}	——	250	250–425	15–25
{Chloride}	——	450	500–650	30–60
(Mixture)				

Substances forming silver salts less soluble than the halides under consideration (*e.g.*, iodide, mercaptides, and sulfides) interfere. The latter two are readily removed by oxidation; iodide is rarely encountered in petroleum. Although it is possible to titrate sulfide, mercaptide, and chloride successively under proper conditions, the amount of nitric acid specified in the procedure here oxidizes the sulfur compounds rather rapidly, and makes the starting point of the chloride titration uncertain. Consequently they should be removed before starting the titration.

As silver mercaptides are of the same order of solubility as silver bromide, the two cannot be distinguished potentiometrically. Any appreciable amount of mercaptans (thiols) in the solution would be evidenced by their odor; their initial potential is the same as that of bromide. Silver mercaptides range in color from white to pale yellow, and often separate as a sticky precipitate that coats the electrodes; they reach equilibrium very slowly when titrated with silver nitrate.

If sulfide is present, the silver electrode is blackened immediately, and the initial potential is off the scale (less than zero). Iodide, if present, starts and breaks off the *p*H scale, and must be titrated on the negative millivolt scale; a well-defined break occurs between iodide and bromide. The procedure here cannot be applied to iodide, however, as it is rapidly oxidized to iodine by nitric acid.

Sulfide and mercaptides may be formed in sodium dehalogenations (see Section 2); and are often present in extracts of crude oils (see Section 3 *c*). They are not found in solutions prepared by the other decomposition procedures given earlier in the chapter.

PROCEDURE

1. Turn on the *p*H meter, allow it to warm up, and make other preliminary adjustments according to the manufacturer's operating instructions. Attach the glass electrode, and a clean silver electrode to the instrument using the appropriate terminal connections. The titration stand should be equipped with an electric motor for continuous stirring, and an off-set burette for making the titration.

NOTES: Practically any *p*H meter can be used in this procedure, and as details of operation differ, the original instructions should be followed.

Glass electrodes should be soaked, or otherwise pretreated according to instructions supplied by their manufacturer.

As remarked in the discussion, it is best to use a clean electrode for this determination.

2. The halide solution prepared in one of the previous sections may contain chloride, or bromide, or both. If only one of these is present the solution should contain a 5-ml excess of 1:1 HNO_3; if both are present 10 ml should be present. When both elements are to be titrated, the potential breaks are sharper if the solution is chilled in an ice bath before titrating.

3. Place the prepared solution in a 400-ml beaker on the titration stand, immerse the electrodes, and set the selector switch on "*p*H." If the initial potential is 200 mv, or less, bromide is indicated. Start the stirring motor, and add standard silver nitrate slowly until the potential reaches about 250 mv. From this point add the standard solution in 0.1-ml increments, recording the corresponding potentials.

NOTES: The volume of the solution should be about 200 ml when the titration is started.

Equilibrium is reached rather slowly near the equivalence point, because bromide ion is strongly adsorbed by silver bromide, and time should be allowed between increments for the system to stabilize.

4. When the first break has occurred, add the solution more rapidly until a potential of about 450 mv is reached, and again add 0.1-ml increments, recording the potentials. After the second break occurs take another reading, or so, then discard the solution and wash the electrodes with water.

> NOTES: If this first break is greater than about 90 mv, there is not likely to be any chloride in the solution (see Table 11:2).
> The glass electrode should be kept in water when not in use.

5. The exact titrations may be computed by the so-called second derivative method.

TYPICAL TITRATION

Milliliters of $AgNO_3$	Potential
0	180
7.3	277
7.4	297
7.5	321
7.6	335
10.2	482
10.3	506
10.4	548
10.5	588

First group: 277 to 297 → 20, 297 to 321 → 24, 321 to 335 → 14; differences 4 and 10.

Second group: 482 to 506 → 24, 506 to 548 → 42, 548 to 588 → 40; differences 18 and 2.

Bromide break: $7.4 + 0.1 \times \dfrac{4}{4+10} = 7.43$ ml

Chloride break: $10.3 + 0.1 \times \dfrac{18}{18+2} = 10.39$ ml

Bromide titration $= 7.43$ ml

Chloride titration $= (10.39 - 7.43) = 2.96$ ml

CALCULATIONS

$$\% \text{ Bromine} = \frac{\text{Br titr} \times N \times 79.92 \times 100}{\text{sample wt} \times 1000}$$

$$\% \text{ Chlorine} = \frac{\text{Cl titr} \times N \times 35.46 \times 100}{\text{sample wt} \times 1000}$$

Halides in gasoline are customarily expressed in grams of halide per gallon, at 60°F. If 100 ml of gasoline is measured at t°C the calculations are as follows:

Bromine, grams/gallon $= 3.025 \times$ titr $\times N(1 + (t - 15.6) \times 10^{-3})$.

Chlorine, grams/gallon $= 1.342 \times$ titr $\times N(1 + (t - 15.6) \times 10^{-3})$.

If the mol ratio is required it is obtained as follows:

$$\text{mol ratio} = \frac{\text{g Cl/gal}}{\text{g Br/gal}} \times 2.25$$

6. Colorimetric Determination of Chloride

The colorimetric determination of chloride with mercuric chloranilate, as described by Barney and Bertolacini,[15] is conveniently applied after the combustion of the sample by the Wickbold method in Section 4 b). The solution obtained by this procedure is free from interfering cations, and as sulfate is the only anion likely to be formed in burning naphthas, the determination can be carried out directly on the absorber solution. Bromide, iodide, flouride, and phosphate interfere, but none of these is likely to occur in either the straight-run, or cracked naphthas used for catalytic reforming feed stock. If this method is to be applied to solutions containing cations other than ammonium ion, the solution must first be passed through a cation exchange column.[15]

The basis of the procedure is the displacement of mercury from its chloranilate salt by chloride ion through the formation of the weak salt, mercuric chloride, with the concomitant release of reddish-purple chloranilic acid anion in the solution at a concentration proportional to the amount of chloride present.

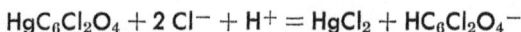

$$HgC_6Cl_2O_4 + 2\ Cl^- + H^+ = HgCl_2 + HC_6Cl_2O_4{}^-$$

The presence of nitric acid increases the intensity of the color, and methyl Cellosolve * is used to decrease the solubility of mercuric chloranilate, and also to suppress the dissociation of mercuric chloride. The maximum absorption in the visible region is at 530 mμ. In a subsequent paper [16] the same authors report greater sensitivity by measuring the absorption in the ultraviolet at 305 mμ.

* Trademark for ethylene glycol monomethyl ether.

As the presence of a small amount of sodium ion is necessary to prevent loss of chloride during evaporation, a special procedure is used for the standardization to compensate for the effect of this cation. To prepare the calibration curve several portions of a standard solution with a known concentration of chlorine is burned and absorbed as in Section 4, *b*). This solution may be prepared with any organic chlorine compound that can be weighed and dissolved conveniently; *o*-chlorotoluene in methyl isobutyl ketone is satisfactory.

PREPARATION

1. REAGENTS:

o-Chlorotoluene: Eastman No. 73; theoretical chlorine content, 28.009%. The compound may be assayed, if desired, by the sodium biphenyl procedure in Section 3, *a*).

Mercuric Chloranilate (2,5-dichloro-3,6-dihydroxy-*p*-benzoquinone mercuric salt): Fisher Scientific Company, Catalog No. M-283.

Methyl Cellosolve: Carbide and Carbon Chemicals Corporation.

Methyl Isobutyl Ketone: Shell Chemical Company.

Nitric Acid, 1N: Dilute 63 ml of concentrated HNO_3 to 1000 ml.

Standard Chlorine Solution: Accurately weigh about 0.36 gram of *o*-chlorotoluene, and transfer to a one-liter volumetric flask containing about 500 ml of methyl isobutyl ketone, dilute to volume with the same solvent, and mix thoroughly.

mg Cl/ml = wt. of *o*-chlorotoluene × 0.280

2. STANDARDIZATION:

1. Burn, by the procedure in Section 4 *b*), the volumes of standard chlorine solution which will yield 1.0, 2.0, 3.0, 4.0, and 5.0 mg of chlorine, and also burn 50 ml of methyl isobutyl ketone as a blank. Transfer the absorber solutions to Erlenmeyer flasks, and evaporate to a volume of about 25 ml.

2. Transfer the evaporated solutions to 100-ml volumetric

flasks, diluting in the process to about 40 ml. To each flask add 5 ml of $1N$ HNO_3 and 50 ml of methyl Cellosolve, dilute to the mark with water, and mix.

NOTE: Heat is evolved when the methyl Cellosolve is added, and the solutions must be cooled before the final dilution.

3. Add 0.2 gram of mercuric chloranilate to each flask, stopper, and shake intermittently for 15 minutes. Filter the solutions through Whatman No. 42 paper, and read the transmittances at 530 mμ using 13-mm cuvettes. Plot the transmittances against the corresponding milligrams of chlorine on semi-logarithmic graph paper.

NOTES: A calibrated scoop is convenient for adding the reagent. The color is stable for several hours.
A second calibration curve may be prepared, using 50-mm cuvettes to cover the 0.1–0.6 mg range, if desired.

PROCEDURE

Evaporate the absorber solution obtained in Section 4 *b*) to a volume of about 25 ml, and proceed as under *Standardization*, starting with step 2. Determine the number of milligrams in the sample from the prepared calibration curve.

7. Volumetric Determination of Fluoride

After organic fluorides have been converted to the ionic form by the Wickbold combustion procedure in Section 4 *a*), the fluoride can be determined by the well-known thorium nitrate titration in a buffered solution, using sodium alizarin sulfonate (Alizarin Red S) as indicator. This method, which has been the subject of many investigations, is inherently accurate, and the ability to judge the end point is readily acquired. It is important, however, to titrate sample solutions under the same conditions as prevail in standardizing the thorium nitrate solution. The significant variables include the concentration of indicator, the volume titrated, pH, and interfering substances, notably sulfate. Although sulfate produces high results the amounts of sulfur in the hydrocarbon streams under consideration do not cause any difficulty.

Matuszak and Brown [18] have applied the method to the

determination of fluorine in hydrocarbons from the hydrofluoric acid alkylation process after decomposition by lamp combustion. They point out that hydrogen peroxide is unnecessary in the absorber, and undesirable, as it decolorizes the indicator. Willard and Horton [19] have studied a large number of indicators for the thorium nitrate titration and give optimum conditions for the use of them.

Most difficulties are caused by the buffer and pH variations. It has been suggested that monochloracetic acid decomposes to yield free chlorine which oxidizes the indicator, thus decreasing its effective concentration and causing high results. Monochloracetic acid is rather easily hydrolyzed to glycolic acid, and this also may be a factor in buffer failure. Refrigerating the solution increases its stability, and the buffer is usable for at least two weeks. Various buffer compositions have been recommended, but the one prescribed for the procedure given here is $0.37M$ in monochloracetic acid, and $0.63M$ in sodium monochloracetate producing a pH of 3.

PREPARATION

1. REAGENTS:

    Buffer Solution: Dissolve 9.45 grams of monochloracetic acid in 60 ml of water. Dissolve 2.5 grams of sodium hydroxide in water, add to the monochloracetic acid solution, and dilute to 100 ml. Keep the solution refrigerated when not in use.

    Indicator Solution: Dissolve 0.5 gram of sodium alizarin sulfonate in 1000 ml of water.

    Sodium Fluoride Solution: Accurately weigh about 2.2 grams of sodium fluoride, dissolve it in water, and dilute to 1000 ml in a volumetric flask. Store the solution in a polyethylene container.

    Thorium Nitrate Solution: Dissolve 7.3 grams of $Th(NO_3)_4 \cdot 4H_2O$ in water, and dilute to 1000 ml. This solution is about 0.05 N.

2. STANDARDIZATION OF THORIUM NITRATE SOLUTION:

Transfer 50 ml of dilute NaOH (4g/l) and 125 ml of water to a Coors 3A porcelain casserole. Add 5.0 ml of sodium fluoride solution, 1.0 ml of indicator, and neutralize with 0.5N HCl just to the yellow color of the indicator. Add 1.0 ml of buffer solution, and titrate with the thorium nitrate solution, stirring vigorously to keep the precipitate dispersed until the color changes from yellow to a faint buff pink. Calculate the normality of the thorium nitrate solution as follows:

$$N = \frac{\text{weight of NaF in grams/liter} \times 5.0}{\text{ml thorium nitrate} \times 42.0}$$

PROCEDURE

To the casserole containing the absorbing solution and washings from the Wickbold combustion in Section 4(a) add 1.0 ml of indicator, and neutralize this with 0.5N HCl until the color changes to yellow. Add 1.0 ml of buffer solution, and titrate with standard thorium nitrate, stirring most vigorously, until the color changes from yellow to a faint buff pink.

$$\% \text{ F} = \frac{\text{ml Th(NO}_3)_4 \times N \times 19 \times 100}{\text{sample weight} \times 1000}$$

NOTES: The volume of the absorber solution and washings should be about 175 ml.

A magnetic stirrer may be used if desired. Allowing the precipitate to settle, and observing its tint is an inferior method for judging the end point; the precipitate should be kept dispersed by vigorous stirrings.

Experience is required to judge the end point, and it is advisable to titrate several portions of the fluoride standard solution to improve perception of the color change. If the titration of a sample exceeds 10 ml, a smaller sample should be burned and titrated.

References

1. AM. SOC. TESTING MATERIALS, Philadelphia, Pa., *1955 Book of ASTM Standards*, Part 5, D808–57T.
2. C. S. PENNINGTON, W. J. SAVOURNIN, and A. L. JULIARD, "Determination of Micro Amounts of Organic Chloride in Petroleum Naphtha," Preprints of General Papers, Division of Petroleum Chemistry, American Chemical Society, Vol. 1, No. 1:71, February, 1956.
3. C. D. HURD and W. A. WILKINSON, "Reaction of Sodium with Chlorosulfides," Division of Petroleum Chemistry, American Chemical Society, Symposium on Organic Sulfur Compounds as Related

to Petroleum, p. 83, March 27 to April 1, 1948, San Francisco, California.

4. B. PECHERER, C. M. GAMBRILL, and G. W. WILCOX, "Determination of Bromine and Chlorine in Gasoline," *Anal. Chem.*, 22:311 (1950).

5. L. M. LIGGETT, "Determination of Organic Halogen with Sodium Biphenyl Reagent," *Ibid.*, 26:748 (1954).

6. C. M. BLAIR, "Determination of Inorganic Salts in Crude Oils," *Ind. Eng. Chem., Anal. Ed.*, 10:207 (1938).

7. C. A. NEILSON, J. S. HUME, and B. H. LINCOLN, "Determination of Salts in Crude Oil," *Ibid.*, 14:464 (1942).

8. K. G. STOFFER, "Salt Contents of Crudes," *Oil and Gas Journal*, 56, No. 5:111; No. 6:129; No. 7:133 (1958).

9. F. E. FREY, "Commercial Alkylation with Hydrogen Fluoride Catalyst," *Chem. Met. Eng.*, 50, No. 11:126 (1943).

10. R. WICKBOLD, "Die Quantitative Verbrennung Fluor-haltiger organischer Substanzen," *Angew. Chem.*, 66:173 (1954).

11. P. B. SWEETSER, "Decomposition of Organic Fluorine Compounds by Wickbold Oxyhydrogen Flame Combustion Method," *Anal. Chem.*, 28:1766 (1956).

12. D. MONNIER, R. VAUCHER, and P. WENGER, "A propos d'un dosage colorimétrique de l'ion fluor," *Helv. Chim. Acta*, 33:1 (1950).

13. R. WICKBOLD, "Neue Schnellmethode zur Halogenbestimmung in organischen Substanzen," *Angew. Chem.*, 64:133 (1952).

14. L. LYKKEN and F. D. TUEMMLER, "Glass Electrode as a Reference Electrode in Electrometric Titrations," *Ind. Eng. Chem., Anal. Ed.*, 14:67 (1942).

15. J. E. BARNEY, II and R. J. BERTOLACINI, "Colorimetric Determination of Chloride with Mercuric Chloranilate," *Anal. Chem.*, 29:1187 (1957).

16. J. E. BARNEY, II and R. J. BERTOLACINI, "Ultraviolet Spectrophotometric Determination of Sulfate, Chloride, and Fluoride with Chloranilic Acid, *Ibid.*, 30:202 (1958).

17. J. G. BERGMANN and J. SANIK, JR., "Determination of Trace Amounts of Chlorine in Naphtha," *Ibid.*, 29:241 (1957).

18. M. P. MATUSZAK and D. R. BROWN, "Thorium Nitrate Titration of Fluoride," *Ind. Eng. Chem., Anal. Ed.*, 17:100 (1945).

19. H. H. WILLARD and C. A. HORTON, "Indicators for Titration of Fluoride with Thorium," *Anal. Chem.*, 22:1190 (1950).

Chapter 12

IRON

Iron is a common oil-soluble contaminant of crude oils, and the use of hematite (Fe_2O_3) for weighting drilling muds, as well as incidental corrosion, leads to its presence in some form in practically all samples. The concentration range of oil-soluble forms is rather wide: from as little as 0.1 ppm to a maximum of about 60 ppm. In common with other elements of the first transition series iron is present to some extent in a volatile form, and is consequently found in heavy distillates. It is somewhat less objectionable than the others, especially copper and nickel, but its determination in feed stocks for catalytic cracking is nevertheless of some importance, and this is covered in Section 1.

The determination of iron in used lubricating oils is often made to evaluate engine performance tests; it is used as an index of corrosion and wear. This application is considered in Section 2. Iron is not used as an additive in petroleum, but iron naphthenate finds some application as an automotive paint drier, and determination in these products is described in Section 3.

From the many different methods that may be used to determine iron, two have been selected which are applicable to most petroleum products. For small amounts the colorimetric procedure using potassium thiocyanate is given in Section 4; and for larger amounts the Zimmermann-Reinhardt volumetric method is prescribed in Section 5.

Two spectrochemical methods for determining iron in used lubricating oils are of interest: Barney and Kimball,[1] and Hansen et al.,[2] have described the use of the emission spectrograph for this purpose; the latter authors report erratic results with oils containing organophosphate additives. Parks

146

and Lykken [3] have proposed an amperometric titration procedure for both iron and copper in oils, which is convenient if equipment is available.

1. Determination of Iron in Distillates

Iron in heavy distillates occurs in the 0.2–6 ppm range, and is partially volatile (see Chapter 2, Tables 2:4 and 2:5). It is therefore necessary to use a fairly large sample, and to pretreat it with a few drops of fuming sulfuric acid to ensure complete recovery. The procedure below is also applicable to crude oils, and to residual fuel oils. Volatility is a consideration in determining iron in crudes, but is not significant in residual fuels. If the direct ashing procedure is used, however, the presence of such elements as sodium and vanadium in the latter, produces fused ashes. The efficacy of sulfuric acid pretreatment, in preventing fused ashes when fluxing agents are present, has been mentioned in Chapter 7, Section 1 a), and it is best to utilize this treatment when determining iron in fuel oils. Its concentration in these products is in the range of 30–50 ppm.

As relatively small amounts of iron are present in any of the three types of oil mentioned above, it is appropriate to use a colorimetric procedure for the final determination. In spite of certain disadvantages, the potassium thiocyanate method is the most convenient; the ratios of iron to interfering elements are particularly favorable for its application without making separations. Interfering elements likely to be present, and their effects, are discussed in Section 4.

A substantial proportion of iron present in any of these products is likely to be in an insoluble form, and in some instances it may be desirable to determine the element in a filtered sample (see Chapter 2, Table 2:6). In any event special care is required to obtain a representative sample. This is particularly important with samples of distillate that have been in storage, as these tend to form sludge. All samples should be warmed to decrease their viscosity, then shaken vigorously, and poured rapidly to ensure a homogeneous sample.

The amount of the sample should be sufficient to yield

.02 to 0.5 mg of iron. In most instances a 200-gram sample is satisfactory for distillates, a 100-gram sample for most crude oils, and 25–50 grams for residual fuels. Tar, asphalt, and other residua contain much heavy material that is not destroyed during the initial burning, and large samples are extremely difficult to handle by wet-oxidation. Fortunately 25 grams of these materials usually produce enough iron for the colorimetric determination in Section 4. Ferric chloride is sometimes used as a catalyst in asphalt manufacture, to shorten the duration of runs, and produce a smoother product. If iron is to be determined in finished asphalt, a 1-gram sample may be wet-oxidized directly, and the solution treated by the procedure in Section 4.

PROCEDURE

1. Heat the sample to reduce its viscosity, and shake it thoroughly. Transfer a suitable amount of it to a clean dry 400-ml beaker. Warm the sample on the steam plate, add 10 drops of fuming sulfuric acid, stir thoroughly, and heat for an additional 10-minute period. Remove the stirring rod, wipe it with a piece of filter paper, and drop the paper into the beaker.

> NOTES: As noted in the discussion, a 200-g sample is used for distillates, a 100-g sample for most crudes, and 25–50 grams for residual fuel oils.
> For 200-gram samples use 8 drops of fuming sulfuric acid; for 25–50-gram samples use 5 drops.

2. Place the beaker in a muffle can (see Chapter 2, Section 2), and ignite the sample with a gas flame. As the sample burns down it may be necessary to apply the flame to maintain burning, but do not heat the bottom of the can directly. Remove the beaker from the can and burn the loose carbon from the upper walls. Cool, add 15–20 ml of H_2SO_4, and digest the mixture on the hot plate for 15–20 minutes. Oxidize as usual with HNO_3, finishing with 30% H_2O_2 and $HClO_4$. The final volume should be about 5 ml.

> NOTE: It is recommended that reference be made to Chapter 2, Section 3 before starting the wet oxidation.

3. Cool the acid solution, dilute it to about 50 ml with water, and boil for about 5 minutes. Cool, filter if necessary, receiving the filtrate in a 100-ml volumetric flask, and treat the solution by the procedure in Section 4.

> NOTES: During the prolonged fuming required to oxidize these materials anhydrous ferric sulfate will separate, and thorough boiling is required to redissolve it. This is an important step in any iron determination that involves fuming with sulfuric acid.
> It is assumed that interfering elements are negligible; see Section 4, for a discussion of these elements, and Section 2 for separations.

2. Determination of Iron in Used Lubricating Oils

The presence of iron in crankcase oils is evidence of corrosion and wear, the amount of iron being more or less proportional to the severity of the condition. As iron is present in used oils in a wide range of concentration, the method of determination to be applied depends upon the amount that can be conveniently isolated. The relatively high levels of compounding used in many lubricating oils limits the sample to 20–30 grams, and in addition, a number of interfering elements must be considered. It is assumed that the following elements may be present, aluminum, barium, calcium, copper, iron, lead, molybdenum, phosphorus, silicon, sodium, and zinc. Aluminum, copper, iron, and lead may be introduced as wear metals; barium, calcium, molybdenum, phosphorus, and zinc are used in compounding; silicon and sodium are occasionally present from contamination by dust.

The first consideration is the procedure for decomposing the sample. As most of the iron is in suspension, or otherwise insoluble, there is no problem of volatility, and all the iron can be retained in a direct ashing procedure. There may be present, however, several elements which cause difficulties if this course is followed. Furthermore, it is mechanically impractical to dissolve the resultant ferric oxide from a bulky, insoluble, and possibly fused ash. Most samples contain too little iron to make feasible a direct wet oxidation, and it is therefore necessary to resort to the method of soft ashing and wet oxidation.

The sample is pretreated with fuming sulfuric acid here also, but for a different reason from that in Section 1. It is a well-known fact that when barium sulfate is precipitated in the presence of ferric salts much iron, very firmly bound, is coprecipitated. Feigl [4] has pointed out that when calcium salts are also present the induced precipitation of calcium sulfate is selective, and no iron is found in the barium sulfate precipitate. The pretreatment with fuming sulfuric acid has a similar effect: barium is precipitated from the oil sample as barium sulfate, and after the subsequent wet oxidation, ferric sulfate dissolves completely when the diluted acid solution is boiled.

The solution is next filtered, and the precipitate, which contains practically all of any barium, lead, or silicon originally present, is discarded. The filtrate contains all of the aluminum, iron, molybdenum, and sodium, and most of the calcium, copper, and zinc. Part of the calcium is carried away with barium sulfate as noted above, and, if the sulfuric acid fuming was prolonged, some of it may have been converted to the insoluble anhydrous calcium sulfate. Copper and zinc are partially coprecipitated if lead sulfate is present.

Two elements that may remain in the solution will interfere in both the colorimetric procedure in Section 4, and the volumetric procedure in Section 5: these are copper and molybdenum. By treating the filtrate with hydrogen sulfide virtually all of the copper is precipitated, and much of the molybdenum; these sulfides are filtered and discarded. The remainder of the molybdenum is separated by oxidizing the iron and precipitating hydrous ferric oxide with ammonium hydroxide. This operation also permits a visual estimate of the amount of iron present, and the selection of the appropriate method for determining it.

PROCEDURE

1. Transfer a 20-gram portion of the homogenized oil sample to a clean dry 400-ml beaker. Warm on the steam plate, add 5 drops of fuming sulfuric acid, and continue to heat for 10 minutes with occasional stirring. Remove the

stirring rod, wipe it with a piece of filter paper, and drop the paper into the beaker. Transfer the beaker to a muffle can, and ignite the oil with a flame. When the fire is extinguished, burn the loose carbon from the upper walls; avoid heating the bottom directly.

2. Cool, add about 20 ml of H_2SO_4, and heat strongly for about 15 minutes on a hot plate. Oxidize the carbonaceous material in the usual way, finishing the oxidation with perchloric acid. The final volume should not exceed 5 ml.

3. Cool the acid solution, dilute to about 100 ml with water, and boil 5 minutes. Sift 0.2 g of $NH_2OH \cdot HCl$ into the boiling solution, and continue the boiling for 5 minutes longer (Note 1). Cool the solution and filter off and discard any insoluble material, using Whatman No. 42 paper. Heat the filtrate to boiling and bubble a rapid stream of H_2S into the hot solution for 5 minutes. Remove the bubbling tube and digest the mixture on the steam plate for 15 minutes. Filter the solution through Whatman No. 40 paper, washing with 1:100 H_2SO_4 saturated with H_2S, and discard the paper and precipitate.

NOTE: Hydroxylamine is added to reduce ferric ion and thus prevent the separation of difficultly filterable colloidal sulfur when the solution is treated with hydrogen sulfide.

4. Transfer the beaker to a hot plate and boil off the H_2S. Cool the solution somewhat, add 1 gram of $(NH_4)_2S_2O_8$, and boil for 10 minutes. Neutralize the solution with NH_4OH until a permanent precipitate forms, then add 10 ml excess. Add a little paper pulp, stir thoroughly, and digest it on the steam plate for 15–30 minutes.

NOTES: Ammonium persulfate is added to oxidize ferrous ion to the ferric state.

Excess ammonia increases the coprecipitation of zinc, but this element is without effect in the subsequent procedures.

5. Filter the solution through Whatman No. 41 paper, and wash the precipitate and beaker thoroughly with water. If the amount of precipitate is judged to contain less than 0.5 mg of iron, place a 100-ml volumetric flask under the funnel, and pour a warm solution containing 10 ml of 1:1 H_2SO_4 and 20 ml of water over the paper and precipitate.

Wash the paper with a small amount of water, and treat the solution by the procedure in Section 4.

NOTE: If more than 0.5 mg of iron has been collected the precipitate can be dissolved, the solution diluted to volume, and an aliquot treated by the procedure in Section 4.

6. If the precipitate is fairly voluminous, place the original beaker under the funnel, and dissolve the precipitate by pouring 25 ml of hot 1:1 HCl over the paper, finally rinsing with water. Treat this solution by the procedure in Section 5.

3. Determination of Iron in Naphthenate Driers

The manufacture of iron naphthenate involves several difficulties, including the necessity for an inert atmosphere, a close control of pH, and the prevention of hydrolysis during dehydration. In a satisfactory product the iron is in the ferrous state; the presence of ferric iron leads to the deposition in storage of a sediment of very finely divided hydrous ferric oxide. As some of this material is invariably present, it is usually desirable to determine the iron content in a filtered or centrifuged sample of the naphthenate.

These products may contain from four to nine per cent of iron, and the special equipment required for manufacturing them precludes any significant contamination by other elements. The organic material may be destroyed by direct ashing, or by fusion with pyrosulfate; the latter is the more rapid method. (For details of the procedure with pyrosulfate fusion, see Chapter 24, *Zinc*.)

Mention was made in Chapter 2, Section 1, of the tendency of mixtures of heavy and light materials to spatter, and exhibit surface turbulence during burning. Naphthenate driers are an example of this, and it is advisable to evaporate the thinner on the steam plate before igniting the weighed sample. The final ignition should be made at about 550°C to avoid the formation of slowly soluble ferric oxide; porcelain dishes are used for the ashing.

It should be mentioned that direct wet oxidation is unsatisfactory for iron naphthenate. The amount of iron handled, 40–90 mg, corresponds to 140–320 mg of ferric sulfate. This salt separates almost completely from fuming sulfuric acid, causes severe bumping and spattering, and makes complete oxidation very difficult and uncertain.

PROCEDURE

1. Weigh and transfer a 1-gram sample of filtered or centrifuged iron naphthenate to a Coors 2/0A porcelain dish and heat this on the steam plate until all the thinner has volatilized. Ignite the sample with a soft flame, and when all of the volatile material has burned off, transfer the dish to a furnace, and ignite it for about 15 minutes at 550°C.

> NOTES: Pyrosulfate fusion requires more working time, but it has advantages over direct ashing. If this procedure is to be used follow the directions in Chapter 24, Section 1.
> The temperature of ignition should not exceed 550°C, or an inordinately long time will be required to dissolve the ferric oxide.

2. Remove the dish from the furnace, cool it to room temperature, add 25 ml of 1:1 HCl to the dish, and warm it on the steam plate until the ferric oxide residue is completely dissolved. Treat the solution by the procedure in Section 5.

> NOTE: If the pyrosulfate fusion procedure is used, add 25 ml of 1:1 HCl to the cooled melt, and warm it on the steam plate until it dissolves.

4. Colorimetric Determination of Iron

Many different colorimetric methods are available for the determination of iron, but only one is given here. The advantages of the thiocyanate method that recommend its application to petroleum products include reaction at high acidity, simplicity of reagents and operations, favorable ratios of iron to interfering metallic elements, and general reliability. A thorough discussion of the thiocyanate method is given by Sandell.[5]

The presence of relatively large amounts of sulfuric acid after wet oxidations is often inconvenient in petroleum analysis, but throughout the book an attempt is made to apply methods of determination that are feasible in this medium. Sulfuric acid decreases the sensitivity of the thiocyanate reaction, and the acid concentration must be controlled between 3 and 5 milliliters per 1000 milliliters.

Among the interfering elements are aluminum, chromium, copper, and vanadium. The effects of these metals are shown in Table 12:1. These data were obtained by taking .302 mg of iron in combination with 3, 6, and 9 mg of the interfering elements, and applying the procedure below; the indicated mg of iron in the presence of each element are tabulated. Calcium, nickel, molybdenum, sodium, phosphate, and zinc are without effect up to 9 milligrams; larger amounts of molybdenum cause high results, and larger amounts of zinc decrease the intensity of the color.

TABLE 12:1

EFFECT OF METALLIC ELEMENTS ON THIOCYANATE
DETERMINATION OF IRON

Interfering Element	Milligrams of Iron Indicated *		
	3 mg	6 mg	9 mg
Aluminum	.302	.311	.319
Chromium (III)	.307	.319	.322
Copper	.337	.370	.403
Vanadium (V)	.307	.310	.312

* 0.302 mg of iron actually present.

The effect of copper is rather serious, but the ratios of iron to copper in the products for which this procedure is recommended are such that the copper concentration is small enough to be without effect in most instances (see for example, Chapter 1, Table 1; Chapter 2, Table 2:5; Chapter 3, Table 3:1). If the solution has been prepared by the procedure in Section 2 copper will, of course, have been eliminated.

PREPARATION

1. REAGENTS:

   Potassium Persulfate Solution: Dissolve 2 grams of $K_2S_2O_8$ in 100 ml of water.

Potassium Thiocyanate Solution: Dissolve 300 grams of KSCN in 1000 ml of water.

Standard Iron Solution: Dissolve 0.100 g of pure iron wire in dilute nitric acid (1:3), boil briefly, and dilute to 1000 ml. 1 ml = 0.10 mg Fe.

2. STANDARDIZATION:

Transfer 0.1, 0.2, 0.3, 0.4, and 0.5 mg of iron to 100-ml volumetric flasks, and include a blank. Bring the total volume of each flask to about 50 ml with water, and add 10 ml of 1:1 H_2SO_4. To each flask add 1.0 ml of 2% $K_2S_2O_8$ solution, and mix. Cool, if necessary, and add 10.0 ml of 30% KSCN, dilute to the mark, and mix thoroughly. After about 5 minutes determine the transmittances of the solutions at 480 mµ, using 13-mm cuvettes, with water as the reference solution. Plot the transmittances against the corresponding milligrams of iron on semi-logarithmic paper.

Dilute the standard iron solution tenfold, and prepare a similar calibration curve covering the range of .01–.07 mg of iron, using 50-mm cuvettes.

PROCEDURE

1. Solutions of iron prepared as directed in the procedures of Sections 1 and 2 should contain the equivalent of 10 ml of 1:1 H_2SO_4, with iron in the ferric state.

NOTE: If an aliquot is used, as suggested in Section 2 (Notes), enough 1:1 H_2SO_4 must be added to make a total of 10 ml.

2. If the solution is not already in a 100-ml volumetric flask, transfer it to one, and cool to room temperature (see notes). Add 1.0 ml of 2% $K_2S_2O_8$ solution, and mix well. Add 10.0 ml of 30% KSCN, dilute to the mark with water, and mix thoroughly. After about 5 minutes determine the transmittance of the solution at 480 mµ using the appropriate cuvette for the color intensity obtained. From the prepared calibration curve for the cuvette used determine the mg of iron in the solution, and calculate the concentration of iron in the original sample.

NOTES: The KSCN solution must not be added to warm solutions as the intensity of color is decreased.

The addition of $K_2S_2O_8$ solution stabilizes the color.

If an aliquot was used account must be taken of the dilution factor in the calculation.

5. Volumetric Determination of Iron

Solutions of iron prepared by the procedures of Sections 2 and 3 are in satisfactory condition for the well-known Zimmermann-Reinhardt determination. Interfering elements, notably copper and molybdenum, were removed in Section 2, and are not likely to be present in iron naphthenate, the preparation of which was discussed in Section 3. The procedure given here is carried out in the usual way, but attention is called to the recommendation of Meites and Thomas [6] that a differential titration of stannous and ferrous ion be made. This has several advantages: the reduced solution need not be cooled, the error introduced by oxidation of mercurous chloride is eliminated, and the need for closely controlling the excess of stannous chloride is obviated. Reinhardt's solution must, of course, be used if the titrant is standard permanganate.

Solutions prepared in Sections 2 and 3 contain ferric iron in hydrochloric acid, and are treated directly by the following procedure.

PREPARATION

1. REAGENTS:

   Mercuric Chloride Solution: Saturate 1000 ml of water with $HgCl_2$.

   Reinhardt's Solution: Dissolve 60 grams of $MnSO_4 \cdot 4H_2O$ in 300 ml of water, and add 240 ml of 1:1 H_2SO_4, 360 ml of water, and 120 ml of 85% phosphoric acid.

   Stannous Chloride Solution: Dissolve 50 grams of $SnCl_2 \cdot 2H_2O$ in 100 ml of HCl, and dilute to 1000 ml with water.

PROCEDURE

1. Heat the hydrochloric acid solution of iron prepared in Section 2 or 3 on the steam plate, and reduce the iron

by adding drops of stannous chloride solution until the yellow color disappears; then add one drop in excess. Cool the solution to room temperature, add rapidly, in one portion, 10 ml of saturated mercuric chloride solution, and allow the mixture to stand for 5 minutes.

2. Transfer the solution to a 600-ml beaker containing 400 ml of water, and 25 ml of Reinhardt's solution, and titrate drop by drop with standard $N/10$ potassium permanganate. Record the volume and normality of the permanganate solution used, and calculate the percentage of iron in the original sample.

References

1. J. E. BARNEY II and W. A. KIMBALL, "Determination of Iron in Used Lubricating Oils," *Anal. Chem.*, 24:1548 (1952).
2. J. HANSEN, P. SKIBA, and C. R. HODGKINS, "Determination of Iron in Used Lubricating Oils by Spectrochemical Analysis," *Ibid.*, 23:1362 (1951).
3. T. D. PARKS and L. LYKKEN, "Determination of Copper and Iron in Oils by Amperometric Titration," *Ibid.*, 22:1503 (1950).
4. F. FEIGL, *Chemistry of Specific, Selective, and Sensitive Reactions*, p. 149, Academic Press, Inc., New York (1949).
5. E. B. SANDELL, *Colorimetric Determination of Traces of Metals*, 2nd Edition, p. 363, Interscience, New York (1950).
6. L. MEITES and H. C. THOMAS, *Advanced Analytical Chemistry*, p. 63, McGraw-Hill, New York (1958).

Chapter 13

LEAD

Probably no other metallic element is determined more frequently in the petroleum laboratory than lead. Its principal importance is in the use of tetraethyllead as an antidetonate in gasolines. This determination is routine, and it is probably most often made by the well-known ASTM extraction procedure,[1] although many other methods are available, a number of which have been collected and published in a book by the Ethyl Corporation.[2] As the determination of tetraethyllead has been thoroughly covered elsewhere it will not be considered in this book.

The analysis of lead naphthenates is of some importance in the petroleum industry, and in Section 1 a procedure is given for determining lead in these products. Lead poisons platinum catalysts, and its concentration in reforming charge stocks is of significance in catalytic reforming operations; it does not seriously affect cracking catalysts. The determination of trace amounts in naphthas and similar products is described in Section 2. Lead additives are used in some special lubricating oils, and the metal is often present in used crankcase oils. A special procedure is given in Section 3 which offers several advantages over the usual methods of handling these oils.

Lead is a common petroleum contaminant, being introduced from lead equipment associated with acid-treating systems, and also from tetraethyllead. The metal occurs only in occasional samples of crude oil, and is infrequently determined in these. Emission spectrography has been little used for lead determinations in petroleum, although a procedure

158

is suggested in the book of the Ethyl Corporation cited above.[2]

1. Determination of Lead in Napthenate Driers

The lead content of naphthenate driers ranges from 8 to 32 per cent. Determination of lead is simplified by the absence of interfering elements and by the ease with which the associated organic material is destroyed. This is one of the few instances in petroleum analysis where wet oxidation is rapid, and the relatively large excess of sulfuric acid is an ideal medium for the determination of the element. It is sufficient to filter the insoluble lead sulfate formed in the wet oxidation, wash, dry, and weigh the precipitate.

Ashing procedures are, in general, not satisfactory for determining lead. Platinum ware cannot be used, and as litharge readily combines with silica to form lead silicate, a large amount of the element may be lost to the glaze if porcelain is used. Gottsch and Grodman [3] reported low and erratic results by direct ashing, and proposed an extraction procedure. The procedure given here is somewhat simpler as the oxidation is practically complete with hydrogen peroxide alone. Swift [4] has shown that satisfactory results are obtained by drying the lead sulfate precipitate with alcohol, and this is prescribed in the procedure that follows. For a large volume of work it may be more convenient to place the crucible in a drying oven for a few minutes rather than aspirating it to remove the excess alcohol; either course gives satisfactory results.

Skoog and Focht [5] have determined lead polarographically in linoleate, naphthenate, and octoate driers after dispersing the sample in tenth-molal dodecylamine acetate; the method is not applicable to lead resinates.

PREPARATION

1. REAGENT:

> Sulfuric Acid Wash Solution: Add 5 ml of H_2SO_4 to 900 ml of water. Add 100 ml of Formula 30 alcohol, and mix.

PROCEDURE

1. Weigh a suitable sample in a Lunge weighing bottle, and transfer it to a 400-ml beaker. Add 10 ml of H_2SO_4, swirl the beaker to disperse the sample, and add 10 ml of 30% H_2O_2. Allow it to stand without swirling or heating until the initial vigorous reaction subsides. Heat the mixture on a hot plate, and as it darkens, carefully add peroxide drop by drop until most of the organic material is destroyed. Finally, heat to heavy fumes and complete the oxidation with a few drops of HNO_3. Cool the solution, add 10 ml of H_2O_2, return to the hot plate, and concentrate the sulfuric acid to a volume of about 5 ml.

> NOTES: The amount of sample taken should be sufficient to yield 100–200 mg of $PbSO_4$; 0.3–0.5-gram is usually satisfactory. A Lunge weighing bottle is required as these products are volatile.
> If less sulfuric acid is used, there is more spattering, and more attention is required.
> The heat of dilution raises the temperature enough to start the reaction. There is usually an induction period, however, and external heat must not be applied until the spontaneous reaction subsides.
> Peroxide is added to destroy nitric acid, in the presence of which the solubility of lead sulfate is appreciably increased.

2. Cool the acid solution, add 50 ml of water, heat to boiling, and boil gently for 5 minutes. Remove from the heat and place the beaker in an ice bath. When the solution has cooled somewhat, add 50 ml of Formula 30 alcohol, and allow to stand in the ice bath for 30 minutes.

> NOTE: Lead sulfate tends to form supersaturated solutions; boiling decreases the relative supersaturation.

3. Filter the precipitate on a tared, medium-porosity, fritted-glass filter crucible using the special sulfuric acid wash solution to transfer and wash the precipitate. Finally, wash the precipitate and crucible thoroughly with Formula 30 alcohol, aspirate the excess, and place the crucible in a drying oven for 5–10 minutes. Remove the crucible, cool, reweigh, and calculate the percentage of lead in the sample.

NOTES: Thorough washing with alcohol is necessary to ensure the complete removal of sulfuric acid as the latter is not volatilized in the subsequent drying.

The gravimetric factor for lead from lead sulfate is 0.6832.

2. Determination of Lead in Light Distillates

The determination of very small amounts of lead is of importance in segregating naphtha feed stocks for reforming operations that employ platinum catalysts. Although lead is not as active as arsenic in poisoning these catalysts (as much as 50 ppb can be tolerated for extended periods of operation), its concentration must be controlled for economical operation.

The dithizone method, first introduced by Fischer,[6] leaves nothing to be desired with respect to sensitivity. Scrupulous care is required, however, in the preparation of reagents and glassware, and experience with the method is necessary for reliable results. A complete discussion of dithizone, and its use for the determination of lead is given by Sandell.[7] The procedure here employs the mixed-color method wherein the color of the lead dithizonate is measured in the presence of excess dithizone. When this method is carried out in the range of 9–10.5 pH certain difficulties arise because of the absorption characteristics of the system. Milkey [8] has given a theoretical treatment of the mixed-color system in this pH range. Snyder [9] improved the method by raising the pH to 11.5. This concentrates most of the excess dithizone in the aqueous phase, extends the range of lead concentration that can be handled, and permits a single strong dithizone solution to be used.

Griffing et al.[10] have described procedures for determining trace amounts of lead in gasolines and naphthas by use of bromine to decompose tetraethyllead. This treatment is not suitable for cracked naphthas, however, as much tarry matter and gum are formed.

Samples for the determination of lead must never be drawn in cans because they can be badly contaminated by metal from the container. In general, the preparation of the reagents, and the procedure prescribed here follows that of Sandell.[7]

PREPARATION

1. REAGENTS:

Ammonia-Cyanide-Sulfite Solution: Dilute 350 ml of NH_4OH and 30 ml of 10% KCN to 1000 ml, and add 1.5 g of Na_2SO_3.

Ammonium Citrate Solution: 50 grams in 100 ml of water. Make the solution ammoniacal, and shake with successive portions of dithizone solution until any lead has been extracted, then remove the excess dithizone by shaking with chloroform.

Bromine Solution: Dilute 30 ml of bromine to 100 ml with chloroform.

Chloroform: Baker's Analyzed. Add 10 ml of absolute alcohol per liter as an inhibitor.

Dithizone Solution: Dissolve 50 mg of dithizone (diphenylthiocarbazone, Eastman No. 3092) in one liter of inhibited chloroform. Keep refrigerated and protected from light.

Hydroxylamine Hydrochloride Solution: Dissolve 20 g of $NH_2OH \cdot HCl$ in 65 ml of water, add NH_4OH until *m*-cresol purple turns yellow, then add 1 ml of 1% sodium diethyldithiocarbamate. Extract with chloroform until carbamates and excess reagent have been removed, add HCl until the indicator is pink, and dilute to 100 ml with water.

Nitric Acid: Dilute colorless, concentrated HNO_3 (air-blown), 1:100 with water.

Potassium Cyanide Solution: Dissolve 50 g of KCN in 100 ml of water. Shake with small portions of dithizone until lead-free, extract excess reagent with chloroform, and dilute to 500 ml with water.

Standard Lead Solution: Dissolve 0.1599 g of recrystallized $Pb(NO_3)_2$ in 1000 ml of 1:100 HNO_3. 1 ml = .01 mg Pb.

2. STANDARDIZATION:

Calibration curves are prepared by carrying known amounts of lead through Parts C and D of the following procedure. Two different curves should be prepared, one using .01, .02, and .03 mg of Pb in a 15-ml volume (10 ml of dithizone solution and 5 ml of $CHCl_3$), and a second using .01, .02, .03, .04, and .05 mg of Pb in a 25-ml volume (10 ml of dithizone solution and 15 ml of $CHCl_3$). Blanks must be carried through the entire procedure, and all glassware should be washed with dilute HNO_3 before use.

PROCEDURE

For lead concentrations in the range of 10–100 ppb a 100-gram sample is satisfactory. Prepare it as follows:

A. IF THE SAMPLE IS A GASOLINE, THINNER, OR STRAIGHT-RUN NAPHTHA

1. Transfer 100 grams to a 300-ml Erlenmeyer flask, add bromine-chloroform solution until the yellow color persists for 5 minutes, and let it stand for 10 minutes longer.

> NOTES: Bromine effectively decomposes tetraethyl lead, and this treatment should be included when its presence is suspected.
>
> Heat until the bromine is expelled, cool to room temperature, and transfer it to a 250-ml separatory funnel, using 25 ml of 1:100 HNO_3 to effect the transfer. Shake the funnel for two minutes, drain the lower layer into a 250-ml beaker, and repeat the extraction.
>
> To the combined extracts add 1 ml of bromine solution, evaporate to a volume of 10 ml, and proceed as described under *Isolation of Lead*.

B. IF THE SAMPLE IS A CRACKED NAPHTHA

1. Transfer a 100-gram portion of it to a 300-ml glass-stoppered Erlenmeyer flask, add 10 ml of HCl, stopper the flask, and shake it for 5 minutes. Remove the stopper, place the flask on the steam plate, and evaporate to dryness.

> NOTE: The evaporation can be hastened by directing an air jet over the top of the flask.

2. Oxidize any remaining tarry material by adding HNO_3 and 30% H_2O_2 in combination; finally evaporate it to dryness. Add 25 ml of 1:100 HNO_3, warm to dissolve any residue, and treat the solution as described next.

C. To Isolate the Lead from Any Sample

1. Transfer the solution of the sample to a 125-ml separatory funnel and dilute it to about 50 ml with 1:100 HNO_3. Add 15 ml of ammonium citrate solution, 1 ml of NH_2OH-HCl solution, and a few drops of thymol blue indicator. Add 5 ml of KCN solution, and if the solution is not alkaline to thymol blue (greenish-blue to blue, pH 9.0–9.5), add NH_4OH until it is alkaline.

> NOTES: Hydroxylamine prevents iron or copper from oxidizing dithizone to diazone, and also reduces any diazone present.
> Cyanide prevents the extraction of zinc by dithizone.

2. Add 5 ml of dithizone solution, and shake the funnel for 15 seconds. Draw the $CHCl_3$ layer into a 60-ml separatory funnel, and extract the aqueous solution with a second 5-ml portion of dithizone. If this extraction shows the mixed color, extract again and combine the extracts in the 60-ml funnel. Shake the combined extracts with 20 ml of water containing a drop of NH_4OH, draw off the chloroform layer into a second 60-ml separatory funnel, extract the aqueous layer with 1 ml of dithizone and add the latter to the main chloroform extract.

> NOTE: A 5-ml portion of dithizone solution is approximately equivalent to 100 micrograms of lead. Therefore, if more than one extraction is required it will probably be necessary to aliquot the final solution before determining the lead.

3. Shake the combined chloroform extracts with 15 ml of 1:100 HNO_3 for 30 seconds, and transfer the acid layer to a 125-ml separatory funnel. Repeat the extraction with a 10-ml portion of dilute HNO_3, and combine this with the first portion. Remove any droplets of dithizone by shaking with $CHCl_3$, and draw off and discard the latter leaving the bore of the stopcock filled with $CHCl_3$. Dry the stem of the funnel with filter paper or a pipe cleaner.

> NOTE: When the $CHCl_3$ layer is extracted with dilute HNO_3 copper remains in the $CHCl_3$ layer and lead is transferred to the acid layer.

D. To Determine Lead

If more than one extraction with dithizone was required (Step 2), the nitric acid solution should be diluted to volume in a volumetric flask, and an aliquot treated as follows:

1. To the lead solution in 1:100 HNO_3 (or to an aliquot diluted to 25 ml with dilute HNO_3) add 75 ml of ammonia-cyanide-sulfite solution. Add 10.0 ml of dithizone solution, and 5.0, or 15.0 ml of $CHCl_3$, and shake it for one minute.

2. Allow the layers to separate, place a plug of glass wool in the stem of the funnel, and drain the $CHCl_3$ into a clean, dry, 13-mm cuvette, discarding the first few milliliters.

3. Measure the transmittance of the solution at 510 mμ without undue delay, and read the corresponding mg of Pb from the appropriate calibration curve, subtracting a blank, if any.

NOTES: This mixture produces a pH of 11.5, and the cyanide prevents interference from copper, iron, zinc; the sulfite maintains a reducing environment which prevents the formation of diazone.

If the amount of lead is less than .02 mg, the total volume should be 15 ml; if between .02 and .05 mg the volume should be 25 ml.

The chloroform should show a pink color with little or no visible green.

The color gradually fades in the light.

3. Determination of Lead in Lubricating Oils

Lead is introduced as tetraethyllead into crankcase oils by dilution of gasoline; from engine deposits, often as the basic sulfate $PbO \cdot PbSO_4$; and as a metal from wear. As the metal may be present in both soluble and insoluble forms, special care is required in the sampling of used lubricating oils. Other metals that may be present include aluminum, barium, calcium, copper, iron, sodium, zinc, as well as phosphorus, and occasionally molybdenum. The use of the alkaline sulfide treatment outlined in Chapter 2, Section 10, offers two outstanding advantages for the determination of lead: two especially troublesome elements, barium and calcium, are not precipitated by the reagent; and a large enough sample can be treated to permit use of a gravimetric method.

The alkaline sulfide treatment was used in Chapter 10, Section 4, for the separation of copper from used lubricating oils, and the same general procedure is used for lead with a few additional considerations. Of the metals possibly present, copper, lead, and zinc are precipitated completely; iron and molybdenum partially; aluminum, barium, calcium, and sodium are not precipitated if in oil-soluble form. Aluminum is sometimes present as the metal, and barium and silicon may be present as insolubles, usually as barium sulfate and silica. The amount of barium sulfate that is collected with lead sulfate is usually small enough, however, to permit a fairly effective separation by ammonium acetate; the same step also separates any silica.

Studies by Waldbauer et al.[11] have shown that copper and zinc are coprecipitated to some extent with lead sulfate. This is of significance in determining copper which is ordinarily present in small amounts, but the effect on a gravimetric determination of lead as the sulfate is small. Thus, with 410 mg of $PbSO_4$ from 1.4 to 1.8 mg of copper were coprecipitated, and with 287 mg of $PbSO_4$ from 1.3 to 1.5 mg of zinc were coprecipitated. These amounts are ordinarily not significant in petroleum analysis. Iron and molybdenum are without effect in the prescribed procedure.

Lead is used as an additive in a few special oils in concentrations of 0.1–0.5 per cent, and its determination in these products by the ASTM procedure[12] invariably gives low results. This method employs a two-gram sample which is wet-oxidized, and subjected to several treatments including two with hydrogen sulfide. The combined lead sulfide precipitates are finally converted to lead sulfate and weighed. This method fails for small amounts of lead, probably because of incomplete precipitation by hydrogen sulfide. Lundell and Hoffman[13] have pointed out that it is usually necessary to add a drop of ammonium hydroxide to "seed out" lead when precipitating small amounts with hydrogen sulfide. This point has not been explored, but in Table 13:1 the results of six single determinations by the ASTM procedure are tabulated, all of which are too low. The weights of $PbSO_4$ obtained from the two-gram samples specified in the ASTM method were from 2 to 7 mg, whereas in the alkaline sulfide treatment, using ten-gram samples, from 30 to 50 mg of $PbSO_4$ were weighed. As the alkaline sulfide treatment was being investigated duplicate determinations were made, and the results are tabulated to indicate the repeatability of the method. No other metals were present in these samples.

TABLE 13:1

COMPARISON OF METHODS FOR MODERATE AMOUNTS
OF LEAD

Sample	ASTM D810	Alkaline Sulfide		Soft Ash, Wet Oxidation
1	0.155	0.305	0.318	0.31
2	0.112	0.210	0.213	0.22
3	0.065	0.190	0.196	0.19
4	0.235	0.318	0.324	0.33
5	0.115	0.355	0.359	0.37
6	0.163	0.339	0.344	0.37

The results obtained by igniting a 30-gram sample, wet-oxidizing the soft ash, filtering, and weighing lead sulfate are tabulated under *Soft Ash, Wet Oxidation.* This method is satisfactory for the determination of lead in new oils that do not contain other additive metal such as barium or calcium, but special care and considerably more time are required than for the alkaline sulfide procedure. The method is not applicable to used lubricating oils.

PREPARATION

1. REAGENTS:

Mixed Solvent: Mix equal volumes of benzene, acetone, and hexane (or petroleum ether).

Sulfide Reagent: Mix 400 ml of Formula 30 alcohol with 100 ml of NH_4OH, and bubble a rapid stream of H_2S into the solution for 5 minutes. Store in a glass bottle with a rubber stopper.

Sulfuric Acid Wash Solution: Add 5 ml of H_2SO_4 to 900 ml of water. Add 100 ml of Formula 30 alcohol, and mix.

PROCEDURE

1. Weigh a suitable sample, say 20 grams, and transfer it to a 250-ml beaker. Add 50 ml of mixed solvent, and mix thoroughly. Add a small amount of paper pulp, disperse by vigorous stirring, and while continuing to stir, add 25 ml of sulfide reagent. Heat mixture to boiling on the steam plate, continuing the stirring. Set aside until the precipitate settles, and filter this through Whatman No. 41 paper. Wash the beaker and precipitate, first with acetone, then with hexane (or petroleum ether), and discard the filtrate.

NOTES: The pulp is added first so that the sulfides will be precipitated in the interstices of the paper, thus facilitating filtration.
The paper can be fitted to the funnel by wetting it with a thinner.

2. Transfer the paper and precipitate to the original beaker, and heat on the steam plate until no solvent odor remains. Add about 7 ml of H_2SO_4, and oxidize the paper and precipitate in the usual way, completing the oxidation with 30% H_2O_2.

3. Cool the acid solution, add 50 ml of water, heat to boiling, and boil gently for 5 minutes. Remove the beaker from the heat and place it in an ice bath. When the solution has cooled somewhat add 50 ml of Formula 30 alcohol, and allow it to stand in the ice bath for 30 minutes.

NOTE: The solution is boiled to ensure complete solution of aluminum and iron sulfates, and to eliminate supersaturation of lead sulfate.

4. Filter the precipitate on a tared, medium-porosity, fritted-glass filter crucible using the special wash solution to transfer and wash the precipitate. Finally, wash the precipitate and crucible thoroughly with Formula 30 alcohol, aspirate the excess, and place the crucible in a drying oven for 5–10 minutes. Remove the crucible, cool, and reweigh.

NOTE: If the sample is a new oil with no other metal present this weight may be used to calculate the lead content.

5. Wash the precipitate with about four 10-ml portions of 50% NH_4Ac, and then with three 15-ml portions of water.

Return the crucible to the drying oven and dry for 30 minutes. Remove the crucible, cool, reweigh, and calculate the percentage of lead in the sample, assuming the loss in weight to be $PbSO_4$.

NOTE: Washing with ammonium acetate dissolves lead sulfate leaving most of any barium sulfate or silica in the crucible.

References

1. AM. SOC. TESTING MATERIALS, *1955 Book of ASTM Standards*, Part 5, D526–53T.
2. *Tetraethyllead. Analytical Methods for its Determination in Gasoline*, Ethyl Corporation, 100 Park Avenue, New York 17, N. Y. (1957).
3. F. GOTTSCH and B. GRODMAN, "An Extraction Method for the Determination of Metals in Boiled Linseed Oil and Driers," *Proceedings, Am. Soc. Testing Materials*, 40:1206 (1940).
4. E. H. SWIFT, *A System of Chemical Analysis*, pp. 129–131, Prentice-Hall, Inc., New York (1946).
5. D. A. SKOOG and R. L. FOCHT, "Polarographic Analysis of Lead Driers," *Anal. Chem.*, 25:1922 (1953).
6. H. FISCHER, *Z. Angew. Chem.*, 42:1025 (1929).
7. E. B. SANDELL, *Colorimetric Determination of Traces of Metals*, 2nd Ed., pp. 87–112; pp. 388–412, Interscience, New York (1950).
8. R. G. MILKEY, "Absorption Characteristics of the Dithizone Mixed Color System," *Anal. Chem.*, 24:1675 (1952).
9. L. J. SNYDER, "Improved Dithizone Method for Determination of Lead," *Ibid.*, 19:684 (1947).
10. M. E. GRIFFING, A. ROZEK, L. J. SNYDER, and S. R. HENDERSON, "Determination of Trace Amounts of Lead in Gasolines and Naphthas," *Ibid.*, 29:190 (1957).
11. L. WALDBAUER, F. W. ROLF, and H. A. FREDIANI, "Spectrographic Studies of Coprecipitation," *Ind. Eng. Chem., Anal. Ed.*, 13:888 (1941).
12. AM. SOC. TESTING MATERIALS, *1955 Book of ASTM Standards*, Part 5, D810–48.
13. G. E. F. LUNDELL and J. I. HOFFMAN, *Outlines of Methods of Chemical Analysis*, p. 51, John Wiley and Sons, New York (1938).

Chapter 14

MANGANESE

Manganese is of relatively little importance in petroleum, occurring in only a few crude oils, and in these in very small amounts. The range of concentration in eight different crudes in which the element was detectable was from .008 to .07 ppm. The Ethyl Corporation [1] has recently made available an experimental manganese anti-knock additive, methylcyclo-pentadienyl manganese tricarbonyl, $(CH_3C_5H_4)Mn(CO)_3$, that may assume importance in gasoline manufacture. Presumably the manganese can be recovered by the conventional hot acid extraction used for the determination of tetraethyllead. As the only determination of manganese at present of general interest in the petroleum industry is in naphthenate driers, only this application is covered in this chapter.

1. Determination of Manganese in Naphthenate Driers

Manganese in naphthenate driers usually is in the range of four to nine per cent, and the organic material is easily decomposed by wet oxidation. The potentiometric titration with permanganate in the presence of excess pyrophosphate developed by Lingane and Karplus [2] is a convenient and accurate method for the determination of manganese. The acid solution of manganous ion is diluted, an excess of tetra-sodium pyrophosphate is added, the pH is adjusted to about 6, and the solution is titrated potentiometrically with standard permanganate. In the presence of pyrophosphate, manganous ion is oxidized to tri-dihydrogen pyrophosphatomanganiate ion according to the following equation:

170

$$4 \, Mn^{++} + MnO_4^- + 8 \, H^+ + 15 \, H_2P_2O_7^{--} =$$

$$5 \, Mn(H_2P_2O_7)_3^{---} + 4 \, H_2O$$

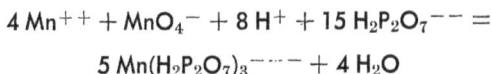

The titration can be made at any pH value between 1 and 8 using a platinum-calomel electrode system; the maximum change in potential, about 180 Mv, is obtained in the pH range of 6–7. The method is remarkably free from interferences; the only ones encountered in the analysis of manganese naphthenate are oxides of nitrogen and hydrogen peroxide.

PROCEDURE

1. Weigh a suitable sample in a Lunge weighing bottle, and transfer it to a 400-ml beaker. Cover the beaker with a ribbed cover glass, and add 10 ml of H_2SO_4, swirling to disperse the sample. Add 10 ml of HNO_3, and when the initial reaction subsides, heat the mixture carefully on a hot plate until the excess HNO_3 evaporates and charring begins, then oxidize as usual with HNO_3 and 30% H_2O_2. When a clean solution is obtained, evaporate the H_2SO_4 to a volume of about 2 ml, adding a little more HNO_3 if the solution darkens. Remove it from the hot plate, cool, and dilute to 250 ml with water.

> NOTES: A convenient amount of manganese is about 80 mg; the amount of sample should be calculated to yield approximately this amount.
> Perchloric acid should not be used as manganese if inconveniently oxidized to manganese dioxide.
> The last oxidizing agent added should be nitric acid; peroxide is not completely destroyed by fuming sulfuric acid, and it consumes permanganate in the subsequent titration.

2. Add 1 gram of sulfamic acid, and mix thoroughly. Add about 25 grams of $Na_4P_2O_7 \cdot 10H_2O$, place the beaker on a titration stand, and adjust the pH to about 6. Titrate the solution with standard $N/10$ $KMnO_4$ using platinum and calomel electrodes, recording the titration and corresponding potentials. Compute the exact titration by the second derivative method (see Chapter 11, Section 5, for an example), and calculate the percentage of manganese.

NOTES: Sulfamic acid is added to remove oxides of nitrogen inevitably left in the acid solution.

If the pH is too high it can be adjusted by adding 1:1 H_2SO_4, and if too low by adding more solid sodium pyrophosphate. The pH can be measured with a pH meter, or visually, using bromthymol blue taken to the yellowish green shade.

When using permanganate, standardized as usual against sodium oxalate, the equivalent weight of manganese is 43.94.

References

1. *Laboratory Use of "Ethyl" Anti-Knock Compound,* Product Service and Safety Section, Ethyl Corporation, 100 Park Avenue, New York 17, N.Y.
2. J. J. LINGANE and R. KARPLUS, "New Method for Determination of Manganese," *Ind. Eng. Chem., Anal. Ed.,* 18:191 (1946).

Chapter 15

MOLYBDENUM

Molybdenum is an element of great importance in petroleum refining, with applications in alloy steels, lubricants, and catalysts. The element rarely occurs in crude oils, but it has been reported in trace amounts in a few isolated samples.[1] Molybdenum is a member of the second transition series, and shows the catalytic activity characteristics of transition elements. Molybdenum catalysts are used in several commercial catalytic processes for desulfurization and reforming. These have been discussed in two articles by Danziger and Milliken.[2] The resistance of these catalysts to poisoning, and their high tolerance for such elements as arsenic, copper, lead, nitrogen, and sulfur make them especially valuable for reforming operations that involve dehydrogenation, dehydrocyclization, and dehydroisomerization. The catalysts selectively hydrogenate nitrogen and sulfur compounds, producing ammonia and hydrogen sulfide. Cobalt is often used with molybdenum in these catalysts; the base is usually aluminum oxide.

Molybdenum disulfide is a useful solid lubricant for certain high-temperature services in which an actual reaction of the lubricant with the metal surface is produced. This "plating" reaction does not occur in internal combustion engines, and in general, lead carbonate is just as effective in greases as molybdenum disulfide. Many so-called "tune-up" oils contain molybdenum additives, such as glycol molybdate, and its presence is always a possibility in used crankcase oils of unknown origins. As the analysis of these oils is complicated by the possible presence of several interfering elements two separate procedures are provided.

Barieau [3] has applied X-ray absorption edge spectrometry

to the determination of molybdenum in samples of hydrocarbons and in solid materials. A large volume of work
would be necessary, however, to justify the initial expense
of the equipment, and for occasional determinations a
chemical method is more practical.

1. Determination of Molybdenum in New Lubricating Oils

Additive molybdenum may occur in concentrations of 0.05
to 0.5 per cent, associated with such elements as barium,
calcium, phosphorus, and sulfur. After destroying the organic material in the wet way, and evaporating most of the
excess of sulfuric acid, molybdenum is determined colorimetrically. The method is based on the addition of thiocyanate to a solution of molybdenum reduced to the pentavalent state by stannous chloride. The red color produced
may be $Mo(CNS)_5$, or complex anions depending upon the
thiocyanate concentration. If this method is used it is best
to extract the red-colored compound with isoamyl alcohol;
the procedure given by Sandell [4] gives satisfactory results
when applied to new oils.

In the procedure prescribed in this section the petroleum
is destroyed by the method of soft ashing and wet oxidation
and as there are ordinarily no interfering elements present,
the molybdenum is reduced in the Jones reductor and then
determined permanganimetrically. Direct ashing cannot be
applied to oils that contain molybdenum because MoO_3 begins to sublime at temperatures around 500°C, and glowing
carbon produces higher temperatures than this even though
the temperature of the furnace is maintained at 500°C. No
losses occur in soft ashing and wet oxidation.

When barium or calcium is present, both may separate
as insoluble sulfates from the fuming sulfuric acid. Neither
exhibits any tendency to carry down molybdenum, however,
and the insoluble material may be filtered and discarded.
The presence of molybdenum is indicated during the wet oxidation by the formation of the red peroxy-acid H_2MoO_8
when peroxide is added. This compound is rapidly decomposed in the hot acid solution, and small amounts may not
be noticeable. Hillebrand et al. [5] point out that the destruc-

tion of organic material with nitric acid may lead to the presence of very stable organic nitrogen compounds that are not destroyed by fuming sulfuric acid. These are reduced in the Jones reductor, and they then consume permanganate, causing high results. They can be destroyed by adding permanganate solution to the sulfuric acid, and heating to fumes.

The presence of these products of decomposition with nitric acid, of which hydroxylamine is one, is indicated by a sliding end point because they are oxidized slowly by permanganate. Common elements that may interfere in the procedure here include chromium, iron, and vanadium. None of these is likely to be encountered in molybdenum additives or their blends, however, and no provision is made for them. If any of them is present the procedure in Section 2 must be followed.

PROCEDURE

1. Transfer a suitable sample (see Notes) to a clean, dry, 400-ml beaker, place the beaker in a muffle can, and ignite the oil with a low flame. When the fire dies out, burn the loose carbon from the upper walls of the beaker taking special care not to heat the bottom or lower part of the beaker directly. Cool it, add 15 ml of H_2SO_4, and fume strongly for about 10 minutes, then oxidize the sample with HNO_3 and H_2O_2 as usual.

> NOTES: For oils containing 0.1 per cent molybdenum, a 50-gram sample is satisfactory. When determining molybdenum in additives about a gram is sufficient, and it should be wet oxidized directly.
>
> It is important to avoid any strong heating because MoO_3 sublimes at low temperature.

2. When a clear solution is obtained, remove the beaker from the heat and cool it. Wash down the cover and sides of the beaker with water, add enough 10% $KMnO_4$ to color the solution permanently, and evaporate to fumes. The final volume of H_2SO_4 should be 2–3 ml. Cool the acid solution, dilute it to about 150 ml, and boil it for 5 minutes. Cool to room temperature, and filter if necessary, using Whatman No. 42 paper.

NOTES: As mentioned in the discussion, this treatment is necessary to destroy organic nitrogen compounds.

When evaporated to this small volume most of any calcium in the solution will separate and it cannot be redissolved.

If barium sulfate is present No. 42 paper is required; separated $CaSO_4$ is retained satisfactorily on No. 40 paper.

3. Prepare the Jones reductor by pouring four or five 25-ml portions of 1:20 H_2SO_4 through the column, discarding the effluent. To the receiver add 25 ml of 10% ferric ammonium sulfate and 5 ml of 85% H_3PO_4, and render the solution just pink with standard $N/10$ $KMnO_4$. Pour the sample solution through the column, receiving the effluent under the surface of the ferric ammonium sulfate solution, and taking care that the sample solution does not fall below the top surface of the zinc column. Wash the column with three 25-ml portions of 1:20 H_2SO_4 followed by a similar amount of water. Titrate the solution in the receiver with standard $N/10$ $KMnO_4$. The titration should be corrected for a reagent blank carried through the procedure.

NOTE: The equivalent weight of molybdenum is 31.98.

2. Determination of Molybdenum in Used Lubricating Oils

The presence in used oils of a number of interfering elements makes the determination of molybdenum by the procedure in Section 1 unsatisfactory. Iron and chromium are both reduced in the Jones reductor, and molybdenum must be separated from them before it is reduced with zinc. As the precipitation of molybdenum by α-benzoinoxime, as described by Knowles,[6] separates the element from all the metals likely to be found in used oils, his procedure has been adopted. A further advantage is that the precipitate can be ignited directly to MoO_3, and weighed.

The advantages of pretreating used oils with fuming sulfuric acid before applying the method of soft ashing and wet oxidation have been mentioned in other chapters. It is particularly applied here to convert lead to infusible lead sulfate, and prevent losses of molybdenum induced by fusion with the beaker. After the wet oxidation step all of any barium, lead, and silicon, and much of any calcium and chromium, will be in insoluble forms virtually uncontaminated by molybdenum. Elements left in solution may include aluminum, calcium, copper, iron, phosphorus, and zinc. When the chilled solution is treated with α-benzoinoxime reagent molybdenum alone is precipitated, and after it is washed, the precipitate is ignited to MoO_3, and weighed. Chromium (VI) and vanadium (V) are also precipitated if present. Chromium is ordinarily reduced with peroxide during wet oxidation; the presence of vanadium in used oils is unlikely. Treatment of the solution with sulfurous acid before precipitation will reduce dichromate and vanadate, and prevent their being precipitated.

PROCEDURE

1. Transfer 20–30 grams of well-mixed sample to a dry 400-ml beaker, add 6–8 drops of fuming sulfuric acid, and heat on the steam plate for about 10 minutes, stirring occasionally. Remove the stirring rod, wipe it with a piece of filter paper, and drop the paper into the beaker. Transfer the beaker to a muffle can, and ignite the oil by heating the can with a flame. When the fire burns out, burn the loose carbon from the upper walls of the beaker, avoiding direct heating of the bottom or lower part. Cool, add 15 ml of H_2SO_4, and fume strongly for about 10 minutes, then oxidize with HNO_3 and H_2O_2 as usual. The final volume of H_2SO_4 should be about 10 ml.

NOTE: The use of perchloric acid may be required when handling badly oxidized oils. The final oxidant, however, should be peroxide as this has the dual function of reducing any dichromate present, and ensuring the removal of nitric acid and its decomposition products. The latter increase the solubility of lead sulfate.

2. Cool the acid solution, dilute it to about 100 ml, and boil it for 5 minutes. Cool it in an ice bath for about 15 minutes, and filter through Whatman No. 42 paper, washing the precipitate with 1:100 H_2SO_4. Dilute the filtrate to 200 ml, chill it in an ice bath, and precipitate molybdenum by adding 10 ml of an alcoholic solution of α-benzoinoxime containing 2 grams per 100 ml of Formula 30 alcohol, and 5 ml extra for each 10 mg of molybdenum. Stir the solution, add

enough bromine water to color the solution a pale yellow, and then add 2–3 ml of reagent. Allow this to stand about 15 minutes with occasional stirring, add a little paper pulp, filter the precipitate on Whatman No. 40 paper, and wash it with 200 ml of a solution containing 5 ml of reagent and 2 ml of H_2SO_4.

> NOTES: The solution is boiled to overcome the tendency of both barium and lead sulfates to form supersaturated solutions.
>
> In an unusual case where chromium and vanadium are present in their highest valency a few milliliters of sulfurous acid should be added here, and the solution boiled until no detectable odor of sulfur dioxide remains.
>
> Twenty milliliters of this reagent will precipitate about seventy-five milligrams of molybdic oxide. A threefold excess should be added over the above theoretical requirement, and not more than 150 mg of molybdenum should be handled.
>
> Bromine is added to counteract a slight reduction of molybdenum by the reagent.

3. Transfer the precipitate and paper to a tared platinum crucible. Dry, and smoke off the paper without burning, using a very low gas flame or an electric radiator, and ignite to constant weight at 500–525°C. From the weight of MoO_3 obtained calculate the percentage of molybdenum in the sample.

> NOTES: A 45 minute ignition is usually sufficient. The MoO_3 should be bright yellow when hot, and pale green when cold.
>
> The gravimetric factor for molybdenum in molybdic oxide is 0.6667.

References

1. M. C. K. JONES and R. L. HARDY, "Petroleum Ash Components and Their Effect on Refractories," *Ind. Eng. Chem.*, 44:2615 (1952).
2. B. H. DANZIGER and J. R. MILLIKEN, "Molybdenum in Petroleum Refining," *Refinery Engineer*, Nov., Dec., 1956.
3. R. E. BARIEAU, "X-ray Absorption Edge Spectrometry as an Analytical Tool," *Anal. Chem.*, 29:348 (1957).
4. E. B. SANDELL, *Colorimetric Determination of Traces of Metals*, 2nd Edition, Interscience, New York (1950).
5. W. F. HILLEBRAND, G. E. F. LUNDELL, J. I. HOFFMAN, and H. A. BRIGHT, *Applied Inorganic Analysis*, 2nd Edition, John Wiley and Sons, New York (1953).
6. H. B. KNOWLES, "The Use of α-Benzoinoxime in the Determination of Molybdenum," *Bur. Standards J. Research*, 9:1 (1932).

Chapter 16

NICKEL

Nickel is probably present to some extent in all crude oils, much of it in a volatile form. As an active dehydrogenation catalyst it alters the selectivity of cracking catalysts, and accurate determination of it in distillate feed stocks is therefore of importance; this application is covered in Section 1. Mixtures of nickel and chromium oleates,[1] and nickel and magnesium soaps,[2] have been proposed as additives for lubricating oil, but they are seldom used in the United States. Nickel organic salts are a popular subject for research, however, and in Section 2 a procedure is provided for handling materials that contain relatively large concentrations of nickel.

1. Determination of Nickel in Distillates

Of the elements commonly found in cuts of heavy gas oil, nickel and copper are the most active dehydrogenation catalysts, and consequently alter the selectivity of cracking catalysts to the greatest extent. The metallic content of feed stock for catalytic cracking is sometimes expressed in "nickel equivalents." The approximate factors for the usual elements are as follows:

Nickel	1
Copper	1
Vanadium	1/10–1/4
Chromium	1/10
Iron	1/50–1/20

Thus, 1 ppm of chromium would have the same effect on the catalyst as 0.1 ppm of nickel or copper.

The presence of the elements named above, as volatile

179

organic compounds in distillates, has been discussed in Chapter 2, Section 2; and of these, nickel is the most volatile. Gamble and Jones [3] have found that approximately 75 per cent of the nickel present may be lost in a direct ashing procedure; similar losses are indicated in Chapter 2, Table 2:4, which shows the effect of sulfuric acid pretreatment on the recovery of metals.

Methods of emission spectrography for determining nickel in charging stocks have been given by Karchmer and Gunn,[4] and by McEvoy et al.[5] The latter apply a novel method of decomposing the sample catalytically by ashing the oil with ground silica-alumina catalyst, and then determining the elements on the catalyst spectrographically. Davis and Hoeck [6] describe the use of x-ray spectrography for determining nickel and vanadium in distillates.

In the procedure given here the sample is pretreated with fuming sulfuric acid, organic material is destroyed by the method of soft ashing and wet oxidation, and nickel is determined colorimetrically by dimethylglyoxime. Interfering elements likely to be present include aluminum, chromium, copper, and iron. Any effect of these or other elements is avoided through isolating the nickel by extraction according to the procedure given by Sandell.[7] This has the further advantage of permitting a 25-ml working volume, which increases the sensitivity.

The addition of dimethylglyoxime to an alkaline solution containing bromine produces with nickel a brown-colored solution. There is doubt as to the composition of the product: it may be a quadrivalent nickel complex, or an oxidized form of dimethylglyoxime. The oxidation potentials of nickel and hypobromite in basic solution, as given by Latimer,[8] indicate that nickel would be oxidized to the quadrivalent state.

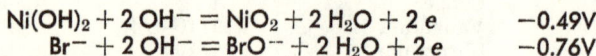

$$Ni(OH)_2 + 2\,OH^- = NiO_2 + 2\,H_2O + 2\,e \qquad -0.49V$$
$$Br^- + 2\,OH^- = BrO^- + 2\,H_2O + 2\,e \qquad -0.76V$$

It should also be noted that if insufficient bromine is present the usual scarlet precipitate of nickelous dimethylglyoxime separates. This eventuality is covered in a note of the procedure given.

PREPARATION

1. REAGENTS:

Bromine Water: Saturated solution.

Dimethylglyoxime Reagent: Dissolve 10 grams of dimethylglyoxime (Eastman No. 98) in 1000 ml of Formula 30 alcohol.

Sodium Citrate Solution: Dissolve 10 grams of $Na_3C_6H_5O_7 \cdot 2H_2O$ in 100 ml of water.

Standard Nickel Solution: Dissolve 0.448 gram of $NiSO_4 \cdot 6H_2O$ in 1000 ml of water containing 3 ml of H_2SO_4. 1 ml = 0.10 mg Ni.

2. STANDARDIZATION:

Dilute the prepared nickel standard tenfold to produce a solution containing .01 mg Ni/ml. Transfer .02, .04, .06, .08, and .10 mg of Ni to a series of 25-ml volumetric flasks, including a blank. Dilute to about 15 ml, add 1 ml of saturated bromine water to each flask, and let them stand for 15 minutes. Add NH_4OH in drops until the yellow color is discharged, then add 1 ml excess. Add 2 ml of 1% dimethylglyoxime solution, and dilute to the mark with water. Mix, let it stand 15 minutes, and determine the transmittances of the solutions of 540 mμ, using 13-mm cuvettes. Plot the transmittances against the corresponding milligrams of nickel on semi-logarithmic paper.

PROCEDURE

1. Heat the submitted sample to reduce its viscosity, and mix thoroughly. Transfer a suitable portion to a clean dry 400-ml beaker, place it on the steam plate, and add 10 drops of fuming H_2SO_4. Stir thoroughly, and heat for about 10 minutes, with occasional stirring. Remove the stirring rod, wipe it with a piece of filter paper, and drop the paper into the beaker.

NOTE: A 200-gram sample is usually satisfactory for distillates.

2. Place the beaker in a muffle can (Chapter 2, Section 2), and reduce the sample to a soft ash as usual. Burn the loose

carbon from the upper part of the beaker, cool it, add 15–20 ml of H_2SO_4, and digest on a hot plate for 15–20 minutes. Oxidize as usual with HNO_3, finishing with 30% H_2O_2, and finally $HClO_4$. Reduce the volume of residual H_2SO_4 to 2–3 ml, cool, dilute to about 25 ml with water, and boil gently for 5 minutes.

> NOTE: Nickel sulfate dissolves rather slowly after being fumed with sulfuric acid; the boiling should not be omitted even if no visible precipitate is present.

3. If insoluble material is present, cool the solution and filter through Whatman No. 42 paper, receiving the filtrate and water washings in a 125-ml separatory funnel. Add a few drops of phenolphthalein indicator, and neutralize with NH_4OH leaving the solution just acidic. Add 10 ml of 10% sodium citrate, cool if necessary, add NH_4OH until the indicator is pink, and then 2–3 drops excess.

> NOTE: If a precipitate forms (usually hydrous ferric oxide) add more sodium citrate.

4. Add 5 ml of 1% dimethylglyoxime solution, and extract with three successive 5-ml portions of $CHCl_3$, shaking the mixture for 30 seconds each time, and combining the $CHCl_3$ extracts in a 60-ml separatory funnel.

5. To the $CHCl_3$ solution add 5-ml of 0.5 N HCl, and shake it for 1 minute. Repeat with a second portion of HCl, transferring the combined acid extracts to a 50-ml beaker. Heat this on the steam plate to evaporate any residual $CHCl_3$, then transfer the remainder to a 25-ml volumetric flask, using a minimum of water. Add 1 ml of saturated bromine water, and let it stand 15 minutes. If the yellow color disappears add more bromine water.

6. Add NH_4OH drop by drop until the yellow color is discharged, then add 1 ml excess. Cool if necessary, add 2 ml of 1% dimethylglyoxime, and dilute to the mark with water. Mix, let it stand 15 minutes, and determine the transmittance at 540 mμ. From the prepared calibration curve determine the milligrams of nickel in the original sample, taking account of any dilution factor.

NOTES: Sometimes the solution does not become completely colorless, but the disappearance of the bromine is easily observed. The reagent should not be added to a warm solution.

If too much nickel is present, a scarlet precipitate will form. If this happens, pour the total contents of the flask into a 100-ml beaker, acidify it with HCl to effect transfer of the precipitate, and evaporate to dryness. Add 5 ml of HCl, and again evaporate to dryness. Take up the residue in a little water, transfer it to a 50-ml volumetric flask, and dilute to the mark. Take an aliquot of this solution, transfer it to a 25-ml volumetric flask, and proceed as before, starting with the addition of bromine water.

2. Gravimetric Determination of Nickel

Nickel additives are infrequently used in petroleum refining, but a few organic salts such as naphthenates, oleates, tallates, and the like are occasionally encountered; these are best handled by a gravimetric procedure. In addition, a few special materials contain enough nickel to warrant a gravimetric determination: e.g., Gilsonite (250 ppm); some Venezuelan crude oils (150–200 ppm); and residual fuel oils (50–100 ppm).

The determination of nickel in organic salts usually presents no difficulties as interfering elements are seldom present. Residual fuel oils contain more metallic elements than any other product likely to be encountered, and the application to these products of the gravimetric method with dimethylglyoxime will be considered. Table 3:1 in Chapter 3, shows the concentration ranges of the metals normally present; they include aluminum, calcium, copper, iron, magnesium, nickel, silicon, sodium, and vanadium. A spectrographic study of the nickel dimethylglyoxime precipitate by Griffing et al.[9] indicates that separation from calcium, magnesium, sodium, and zinc is complete; aluminum, chromium, copper, iron, and manganese coprecipitate. The addition of tartrate prevents contamination of the precipitate by aluminum, chromium, and iron, but the separation from significant amounts of copper is not satisfactory. In the case of fuel oils the amount of copper normally present (0.2–5 ppm) is insignificant. If the element is present in larger amounts it must be removed by treatment with hydrogen sulfide before nickel is precipitated.

Table 16:1 shows the results obtained by three methods on four separate samples of fuel oil. The colorimetric method described in Section 1 was applied after direct wet-oxidation of a 2-gram sample; the gravimetric method given here was applied, using a 100-gram sample with sulfuric acid pretreatment; spectrographic determinations were made by ashing a sample directly on a matrix of strontium sulfate.

TABLE 16:1

COMPARISON OF METHODS FOR MODERATE AMOUNTS OF NICKEL

Residual Fuel Oils	ppm Nickel		
	Colorimetric	Gravimetric	Spectrographic
I	89	80	64
II	76	66	53
III	90	80	62
IV	73	62	56

It will be noted that there is evidence of volatility of nickel even in these heavy residual materials. The lowest results were obtained by the spectrographic method which involves direct ashing; the highest results were obtained by direct wet-oxidation in the colorimetric method. If the colorimetric results are taken as more reliable, the 10–15 per cent lower results from the gravimetric procedure, using sulfuric acid pretreatment, are probably explained by inadequate contact of these very heavy samples with the acid.

The usual precipitation of nickel dimethylglyoxime by ammonia is prescribed in the following procedure, but attention is called to the method of urea hydrolysis of Bickerdike and Willard.[10] Their procedure is especially useful when it is necessary to handle large amounts of nickel (100 mg or more).

PROCEDURE

1. Transfer enough sample to yield 5–25 mg of nickel to a Coors No. 2 porcelain evaporating dish, add a few drops of fuming H_2SO_4, and warm this on a steam plate with occasional stirring for 10 minutes. Wipe the stirring rod with a piece of filter paper, and drop the paper into the dish. Ignite

the oil with a flame, and when the sample has been consumed, burn off the loose carbon with a low flame. Transfer the dish to a furnace, and ignite it at 550°C until all carbon is oxidized. Cool the dish, add 10 ml of HCl, cover, and warm it on the steam plate until the residue dissolves.

> NOTES: This procedure applies to crude oils, residua, lube oil blends, etc. When handling material containing several per cent of the element a small sample should be weighed and wet-oxidized. When a clean solution is obtained proceed with the second paragraph of the procedure.
>
> For a 100-gram sample use 10 drops of sulfuric acid, and proportionately less for smaller samples.
>
> In many instances the entire ash does not dissolve; 15 minutes of heating is sufficient.

2. Dilute to 50-ml with water, boil gently for 5 minutes, filtering and discarding any insoluble residue. Add 2 grams of ammonium tartrate, dilute to 150 ml, heat nearly to boiling, add 20 ml of a 1% alcoholic solution of dimethylglyoxime, neutralize with NH_4OH until a scarlet precipitate forms, and then add a few drops excess.

3. Digest on the steam plate for 1 hour, cool, filter on a tared, medium-porosity, fritted-glass filter crucible, and wash with water. Dry the precipitate in an oven for 45 minutes at 105°C, cool, and reweigh.

> NOTE: The gravimetric factor for nickel from its dimethylglyoximate is 0.2032.

References

1. C. C. WAKEFIELD AND CO., LTD., and EVANS, Brit. Patent 474,156 (1937).
2. STANDARD OIL DEVELOPMENT CO., Brit. Patent 535,777 (1941).
3. L. W. GAMBLE and W. H. JONES, "Determination of Trace Metals in Petroleum," *Anal. Chem.*, 27:1456 (1955).
4. J. H. KARCHMER and E. L. GUNN, "Determination of Trace Metals in Petroleum Fractions," *Ibid.*, 24:1733 (1952).
5. J. E. McEVOY, T. H. MILLIKEN, and A. L. JULIARD, "Spectrographic Determination of Nickel and Vanadium in Petroleum Products by Catalytic Ashing," *Ibid.*, 27:1869 (1955).
6. E. N. DAVIS and B. C. HOECK, "X-Ray Spectrographic Method for the Determination of Vanadium and Nickel in Residual Fuels and Charging Stocks," *Ibid.*, 27:1880 (1955).
7. E. B. SANDELL, *Colorimetric Determination of Traces of Metals*, 2nd Ed., Interscience, New York (1950).

8. W. M. LATIMER, *The Oxidation States of the Elements and Their Potentials in Aqueous Solutions*, Prentice-Hall, New York (1952).
9. M. GRIFFING, T. DE VRIES, and M. G. MELLON, "Spectrographic Analysis of Organic Precipitates," *Anal. Chem.*, 19:654 (1947).
10. E. L. BICKERDIKE and H. H. WILLARD, "Dimethylglyoxime for Determination of Nickel in Large Amounts," *Ibid.*, 24:1026 (1952).

Chapter 17

NITROGEN

The determination of nitrogen and its compounds has become increasingly important in recent years principally because of their undesirable effects in refining operations and in petroleum products. Ball *et al.*[1] have summarized the results of a survey by the Bureau of Mines of 153 United States crude oils, which presents evidence that the ratio of nitrogen to carbon residue is related to the geological origin of the crude. The tendency of burner fuels to form sludges in storage and the formation of gum in diesel fuels have been related to the presence of pyrroles. These compounds are sensitive to hydrolysis and oxidation as well as to heat and light. Additives are available, organic amines for example, which peptize gum and prevent its being precipitated. Dimpfl *et al.*[2] have described methods for evaluating fuel oil additives, and Muhs and Weiss[3] have developed a procedure for determining pyrrolic nitrogen in distillates.

The presence of nitrogen bases (homologs of pyridine, quinoline, and isoquinoline) in distillate feed stocks for catalytic cracking, reforming, isomerization, and polymerization adversely affects the catalysts. They reduce the activity of cracking catalysts by neutralizing the active acidic centers, thus lowering gasoline yields. The effect of nitrogen compounds on platinum-alumina catalysts has been investigated by Meisel *et al.*,[4] who found that both the acidic centers and the platinum are affected. Nitrogen is undesirable in gasoline as it promotes the formation of lacquer on piston rings, and the formation of deposits in carburetors.

The Kjeldahl method has proved reliable for the determination of nitrogen in crude oils, residua, lubricating oils, and additives, and two variations of this procedure are covered in Section 1. Noble [5] has applied the phenol-sodium hypochlorite colorimetric method after Kjeldahl digestion for determining small concentrations of nitrogen; Taras [6] describes several colorimetric methods for nitrogen in a variety of materials.

Because of the deleterious effect of nitrogen bases on various refining catalysts, the determination of total basic nitrogen in various charge stocks is important, and this is described in Section 2. Attention has been given to the possibility of converting nitrogen compounds in petroleum into useful products,[7] some of the more important of which are quaternary ammonium compounds.[8] As these are finding increasing use in the petroleum industry a procedure is given in Section 3 for determining them.

1. Determination of Total Nitrogen by the Kjeldahl Method

The Kjeldahl method for nitrogen is probably used in more different laboratories than any other chemical method, and as a consequence, a great many modifications have been proposed through the years. Kirk [9] has reviewed the various catalysts, conditions for digestion, and other factors, and gives references to much of the early work.

A joint study by the Bureau of Mines and the Union Oil Company of California [10] has shown that the temperature of digestion is the most critical factor in the Kjeldahl method. Data are presented showing that the temperature must be controlled within the rather narrow range of 370–410°C; this is accomplished by using the proper proportions of potassium sulfate and sulfuric acid. McCutchan and Roth [11] investigated the effect of o-mercaptobenzoic acid on nitro compounds, and found that it gives quantitative recovery of nitrogen. Although nitro compounds are unusual in petroleum, the pretreatment of all samples with this reagent is advisable. There is evidence that this treatment also improves recovery of nitrogen in other types of compounds. Table 17:1 shows the results obtained on three chemicals of industrial grade with and without pretreatment with o-mercaptobenzoic acid. It can be seen that without pretreatment the results for nitrogen are low and erratic. This may be because of the nature of the nitrogen impurities in these chemicals of technical grade.

TABLE 17:1

EFFECT OF O-MERCAPTOBENZOIC ACID ON
RECOVERY OF NITROGEN

Compound *	Without Pre-treatment	With Pre-treatment	Theoretical
Trimethyldihydroquinoline	7.04	8.28	8.1
	7.21	8.30	
	7.89		
Diphenylamine	7.49	7.91	8.3
	7.59	7.98	
N-Phenyl-1-naphthylamine	5.81	6.25	6.4
	5.85	6.30	

* These are products of industrial grade; the theoretical percentages for nitrogen apply to the pure compounds.

Lake [12] has summarized the results of a collaborative effort by fourteen laboratories to evaluate the Kjeldahl, Dumas, and ter Meulen methods for nitrogen in petroleum and shale oil. It was concluded that the Kjeldahl procedure is applicable to the majority of petroleum samples provided that the temperature of digestion is controlled and mercury is used as the catalyst. The method determines pyridines, quinolines, isoquinolines, purines, pyrimidines, amines, amides, oximes, carbazoles, and other compounds; it is not suitable for hydrazines, pyrazoles, or azo compounds.

The use of catalysts other than mercury has been recommended, notably selenium and copper. Selenium has fallen somewhat into disrepute, however, as it gives low results in some instances, and copper is likely to do the same, particularly in short digestions.

A. KJELDAHL-WILFARTH-GUNNING PROCEDURE

The method of digestion with sulfuric acid for converting organic nitrogen compounds to ammonium salts was first described by Kjeldahl. The use of mercury as an oxidation catalyst was introduced by Wilfarth,[13] and Gunning[14] first used potassium sulfate to raise the temperature of the digestion mixture. The combined procedure has been a standard method [15] for many years, and nitrogen can be determined accurately in most petroleum products by this method provided the pretreatment with o-mercaptobenzoic acid [11] is used.

The procedure consists of adding a weighed sample to a mixture of sulfuric acid and o-mercaptobenzoic acid, and charring it. Potassium sulfate and mercuric oxide are added, and the temperature is gradually raised until the mixture clears, after which the solution is boiled for an additional hour. The mixture is then cooled, diluted, and neutralized with sodium hydroxide solution. Sodium sulfide solution is added to precipitate mercury, and the ammonia formed in the digestion is steam-distilled into a boric acid solution, and titrated with standard acid using methyl purple as indicator.

A summary by Deal et al.[16] of nitrogen compounds that have been found in petroleum shows a predominance of homologs of pyridine, quinoline, and isoquinoline. Older references [17] indicate that these are particularly refractory, requiring many hours of boiling to decompose them completely. Compounds of these types yield readily to the decomposition procedure given here, however, and they can be satisfactorily determined after a relatively short period of digestion.

The following procedure is applicable to crude oils, various distillates, and residua as well as to many additives, including the various amines used as inhibitors of oxidation and corrosion, and aromatic amines used as metal deactivators and fuel oil additives. The size of a sample may range from 0.1 gram to a maximum of 3.5 grams. Table 17:2 shows appropriate sizes of samples for different nitrogen contents with the corresponding titration volumes. This table applies to both procedures A and B.

TABLE 17:2

SIZES OF SAMPLES FOR KJELDAHL NITROGEN METHOD

Nitrogen, %	Size of Sample, Grams	Titration, ml of 0.1N H⁺
0.01–0.2	3.5	0.25–5 *
0.2–1	2.0	2.8–14
1–5	1.0	7–35
5–10	0.5	18–35
10+	0.1	7+

* Samples in this range should be titrated with 0.01N H⁺.

PREPARATION

1. REAGENTS:

Boric Acid Solution: Dissolve 20 grams of H_3BO_3 in 1000 ml of water.

Kel-Pack Nitrogen Testing Powder Formula No. 3: Each package contains 9 grams of K_2SO_4 and 0.35 gram of HgO; use two packages for each determination. Available from Harshaw Chemical Company, 5321 East 8th St., Oakland, Calif., and many laboratory supply houses.

o-Mercaptobenzoic Acid: Eastman No. T2805.

Methyl Purple Indicator: Prepared aqueous solution available from Fleischer Chemical Co., Benjamin Franklin Station, Washington, D.C.

Sodium Hydroxide Solution, 40% aqueous: 572 grams of NaOH per liter.

Sodium Sulfide Solution: Dissolve 100 grams of $Na_2S \cdot 9H_2O$ in 1000 ml of water.

PROCEDURE

1. Transfer 15 ml of H_2SO_4 to a 500-ml Kjeldahl flask, add 1.0 gram of o-mercaptobenzoic acid, and swirl to mix. Transfer a suitable weight of sample (see Table 17:2) to the flask, then add 10 ml of H_2SO_4 and 5 ml additional for each gram of sample. Place the flask on a suitable stand in a fume hood, and heat moderately until the sample is well charred, then increase the heat until incipient spattering.

NOTES: As some lots of o-mercaptobenzoic acid contain appreciable amounts of nitrogen, the quantity added should be weighed rather closely.

Volatile samples may be weighed in a Lunge weighing bottle; samples of oil are conveniently weighed in Coors 00000 porcelain crucibles. As gelatine capsules contain about 16% nitrogen, they cannot be used.

2. Cool the mixture, and add the contents of two packages of Kel-Pack Formula No. 3. Return the flask to the heat and maintain a moderate temperature until foaming is brought under control, then gradually increase the tempera-

ture until the mixture boils and clears. Boil the solution for 60 minutes after clearing.

NOTES: The mixture must be fairly cool or severe foaming may occur. This is caused by the water evolved in the following sequence:

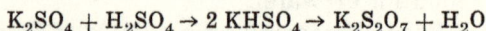

$$K_2SO_4 + H_2SO_4 \rightarrow 2\ KHSO_4 \rightarrow K_2S_2O_7 + H_2O$$

(see Reference 12).
The solution must not be boiled longer than 75 minutes after clearing or ammonia may be lost.

3. Cool the acid solution to room temperature, dilute it to about 250 ml with water, and cool it again to room temperature. Add a few boiling chips to control bumping, and attach a distilling head consisting of a rubber stopper fitted with a dropping funnel and a spray trap, the latter connecting to the top of a vertical Allihn-type condenser. Attach a glass tube to the bottom outlet of the condenser, and place a 500-ml Erlenmeyer flask, containing 25 ml of 2% H_3BO_3 solution under the condenser so that the glass tube dips beneath the surface of the solution.

NOTE: Glass beads, carborundum granules, or carbon chips are satisfactory anti-bumping agents; mossy zinc has also been recommended.

4. Transfer enough 40% NaOH solution to the dropping funnel to neutralize the H_2SO_4 used, and admit the solution slowly while swirling the contents of the flask. Transfer 20 ml of 10% Na_2S solution to the dropping funnel, and drain this solution into the flask. Heat the flask, preferably with a Bunsen burner, until vigorous boiling occurs, and steam-distill the ammonia until about 100 ml of distillate has been collected. Remove the source of heat, vent the system by opening the stopcock on the dropping funnel, disconnect the spray trap from the condenser, and rinse with a little water, collecting the washings in the Erlenmeyer flask.

NOTES: 75 ml of 40% NaOH will neutralize 30 ml of H_2SO_4; less than this is required as H_2SO_4 is consumed in the digestion.
Mercuric ion forms stable ammonobasic salts such as $(HgNH_2)_2SO_4$ with ammonia. These are decomposed upon addition of sulfide because the extremely insoluble HgS is precipitated with concomitant release of NH_3.
If desired, the indicator may be added at the outset, and the ammonia titrated as it is evolved.

5. Add a few drops of methyl purple indicator, and titrate with standard acid of appropriate strength. The change in the indicator color is from green to gray to purple-pink. A blank, consisting of a combination of 1.0 gram of o-mercapto-benzoic acid and 1 gram of sucrose, should be carried through the procedure, and the titration of the sample should be corrected accordingly.

$$\% \, N = \frac{(titr\text{-}blk) \times N \times 14}{sample \; wt \times 10}$$

NOTE: A blank of 0.04–0.06 meq is normal.

B. STUBBLEFIELD-DE TURK PROCEDURE

Some high-molecular-weight polymers used to improve viscosity index in lubricating oils contain moderate amounts of nitrogen which is incompletely recovered by the procedure in Section 1A. These materials can be effectively handled in a method developed by Stubblefield and De Turk,[18] but pre-treatment with o-mercaptobenzoic acid is necessary. The digestion mixture consists of dibasic potassium phosphate, ferric sulfate, and sulfuric acid, with mercury as the oxidation catalyst. As ferric sulfate is insoluble in fuming sulfuric acid, it provides particles around which a film of decomposing sample forms, thus exposing a large surface to the action of sulfur trioxide; it also provides nuclei to facilitate the formation of bubbles. Dibasic potassium phosphate raises the boiling point of the digestion mixture, and prevents bumping in the subsequent distillation. Table 17:3 shows results obtained by various methods on a single sample of a solution of an acrylic resin.

TABLE 17:3

COMPARISON OF METHODS FOR NITROGEN

Procedure	% Nitrogen
1 A Without pretreatment	0.34, 0.36
1 A With pretreatment	0.40, 0.41
1 B Without pretreatment	0.33, 0.37
1 B With pretreatment	0.44, 0.46,
	0.47, 0.47
Dumas Method	0.46, 0.47

PROCEDURE

1. Transfer 15 ml of H_2SO_4 to a 500-ml Kjeldahl flask, add 1.0 gram of o-mercaptobenzoic acid, and swirl to mix. Transfer a weighed sample (see Table 17:2 to the flask, and add 15 ml of H_2SO_4. Place the flask on a suitable stand in a fume hood, and heat moderately until the sample is well charred, then increase the heat until spattering commences.

2. Cool the mixture, add 10 grams of K_2HPO_4, 6 grams of $Fe_2(SO_4)_3$, and 0.6 gram of metallic mercury. Return it to the heat and maintain a moderate temperature until foaming is brought under control, then gradually increase the heat and rotate the flask at frequent intervals until a uniform yellow mass is obtained. Care must be taken not to overheat the mixture as H_2SO_4 is easily lost; if the mass becomes too thick to flow, add a few ml of H_2SO_4. When all carbonaceous material has decomposed and a uniform yellow paste remains the digestion is complete.

> NOTES: This mixture must be fairly cool as considerable heat is generated when the next reagents are added.
> A solution is not obtained in this digestion, but rather a viscous mixture.

3. Cool the mixture and proceed by the third paragraph of the procedure in Section 1A.

2. Determination of Basic Nitrogen

In this section two applications of the acidimetric determination in nonaqueous solvents of basic nitrogen are considered. The first and more important is the determination of nitrogen bases in distillates used as charging stocks for catalytic cracking and for re-forming processes using platinum as catalyst. The poisoning effect of these compounds on the catalysts for these processes has been described by Mills et al.,[19] and by Meisel et al.[4]

The procedure given here, which is essentially that given by Fritz,[20] depends upon the leveling effect of acetic acid on weak bases. Nitrogen bases having aqueous dissociation constants greater than about 10^{-13} can be titrated by a strong acid in acetic acid solution because their basic strength in this solvent is much greater than in water. In this method methyl violet is used as an indicator; others have been described using potentiometric titrations.[16, 21]

Formerly it was postulated that ntirogen bases react with the acidic solvent as represented by the following equation:

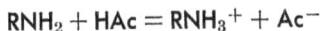

$$RNH_2 + HAc = RNH_3^+ + Ac^-$$

The acetate ion was then titrated by the strong acid used as titrant. It now appears, however, that the formation of RNH_3Ac, a hydrogen-bonded adduct, is more likely. The addition of a stronger acid, e.g. perchloric acid, displaces acetic acid from the hydrogen-bonded adduct. When one drop of the titrant is added in excess the sharp drop in pH, due to the presence of this free acid, is shown by a potentiometer or by the change in color of an indicator. The tendency of heterocyclic basic nitrogen compounds to form insoluble perchlorates in acetic acid medium tends to force the reaction to completion even when the bases are very weak.[20]

The perchloric acid standard solution is prepared by mixing the required amount of 72% $HClO_4$ with glacial acetic acid, and adding a small amount of acetic anhydride to remove the water. Acetic acid, acetic anhydride, and water comprise a fairly inert system unless heated or catalyzed. Perchloric acid is a catalyst for the hydration of acetic anhydride, however, and equilibrium is reached in a relatively short time. The solution may be standardized against anhydrous sodium carbonate or 1,3-diphenylguanidine; the latter has the higher equivalent weight, and gives a sharper end point.

The naphthas used as feed stock for re-forming catalysts do not require dilution, but heavy gas oils used for catalytic cracking must be diluted with a solvent before titration. Suitable solvents include benzene, chlorobenzene, nitrobenzene, isooctane, and methyl isobutyl ketone.

Dreher et al.[22] have described the use of sodium N-octadecylterephthalamate as a thickener for multipurpose lubricating grease. This product is prepared by reacting sodium hydroxide with the methyl ester of N-octadecylterephthalamic acid. The latter is synthesized from dimethylterephthalate and a mixture of octadecyl- and hexadecylamines, and the presence of excess amines is undesirable in the product. As these amines react as bases they can be determined by dissolving a small sample in benzene and titrating with standard perchloric acid, using methyl violet indicator as in the procedure given here.

PREPARATION

1. REAGENTS:
    Acetic Anhydride: Reagent grade.
    1,3-Diphenylguanidine: Eastman No. 1270.
    Glacial Acetic Acid: Reagent grade.
    Methyl Violet Indicator: Dissolve 1.0 gram of methyl violet, Eastman No. 1309, in 100 ml of glacial HAc. The solution is stable.
    Perchloric Acid, 0.1 N: Mix 8.5 ml of 72% $HClO_4$ with 500 ml of glacial HAc, add 20 ml of acetic anhydride, dilute to 1000 ml, and allow to stand overnight.

2. STANDARDIZATION OF PERCHLORIC ACID:

Accurately weigh about 0.2 gram of 1,3-diphenylguanidine, and transfer it to a 500-ml Erlenmeyer flask. Add 150 ml of glacial HAc, and a few drops of methyl violet indicator, and titrate to a blue color with the prepared solution of perchloric acid. Calculate the normality of the perchloric acid solution as follows:

$$N = \frac{W \times 1000}{V \times 211.3}$$

where:

    N = Normality of $HClO_4$
    W = Grams of 1,3-diphenylguanidine
    V = Milliliters of $HClO_4$

For determining small concentrations of nitrogen bases dilute the 0.1N $HClO_4$ tenfold with glacial HAc to produce an 0.01N solution.

PROCEDURE

A. NAPHTHAS

Transfer 100 ml of a suitable solvent (isooctane, benzene, etc.), 100 ml of glacial HAc, and a few drops of methyl violet indicator to a 500-ml Erlenmeyer flask, and titrate with standard $HClO_4$ to the end point (blue color). Record this titration as the blank.

NOTES: The color change is from violet to blue to blue-green.
Either 0.1 N or 0.01 N $HClO_4$ may be used depending upon the concentration of basic nitrogen expected; for naphthas and gas oils 0.01 N $HClO_4$ is usually appropriate.

2. Transfer a suitable sample, 100 ml of glacial HAc, and a few drops of methyl violet indicator to a 500-ml Erlenmeyer flask, and titrate with standard $HClO_4$ to the blue color. Record this titration as the sample titration.

$$\text{ppm basic } N = \frac{(\text{sample titr-blk}) \times N \times 14 \times 10^6}{\text{ml sample} \times \text{sp gr} \times 10^3}$$

NOTE: Using 0.01 N $HClO_4$ the sample size may be selected from the following table:

ppm Basic N	ml Sample
100–200	5
20–100	25
5–20	100
less than 5	200

B. SOLID MATERIALS

Weigh a suitable sample, and dissolve in benzene, warming if necessary to effect solution. Add 100 ml of glacial HAc and a few drops of methyl violet indicator, and titrate with standard $HClO_4$ to the blue color.

$$\% \text{ basic } N = \frac{\text{ml titr} \times N \times 14 \times 100}{\text{sample wt} \times 1000}$$

NOTES: For samples of the order of 0.1% basic nitrogen a 2-gram sample is satisfactory.
With very dark samples where the visual end point cannot be seen the potentiometric procedure must be used. A pH meter equipped with a glass and a sleeve-type calomel electrode is required. The prepared solution is titrated in the usual manner to the maximum change in potential using the millivolt scale (plus).

3. Determination of Quaternary Ammonium Compounds

The preparation of quaternary ammonium salts from dodecylbenzene has been described by Darragh and Stayner.[8] These compounds are used in ore flotation processes, textile finishing, and water treatment. They are surface-active agents with algaecidal and germicidal properties.

Most methods for determining these compounds are based on the formation of insoluble salts with various anions. Hager et al.[23] have devised a procedure based on the insolubility of the triiodides of quaternary compounds, and Flotow [24] has given several other methods, including precipitation as dichromate which is prescribed here.

The molecular weight of the quaternary salt to be determined must be known; those prepared from dodecylbenzene derived from petroleum have molecular weights a little higher than calculated because of the presence of higher homologs. Thus, the calculated molecular weight of N-dodecylbenzyl-N,N,N-trimethylammonium chloride is 354, while the empirical weight is about 358; the calculated molecular weight of N-dodecylbenzyl-N,N,N-triethylammonium chloride is 396, and the empirical weight is about 400. These differences, however, are not significant in the analysis of the commercial products sold as concentrated aqueous solutions. The triethyl compound will be taken as an example in the following discussion of the procedure.

A measured amount of standard potassium dichromate is added to a weighed sample of the quaternary compound, and the insoluble quaternary dichromate precipitates.

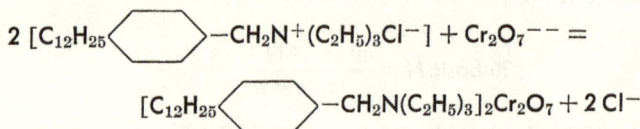

$$2 [C_{12}H_{25}\langle \rangle -CH_2N^+(C_2H_5)_3Cl^-] + Cr_2O_7^{--} =$$

$$[C_{12}H_{25}\langle \rangle -CH_2N(C_2H_5)_3]_2Cr_2O_7 + 2 Cl^-$$

The precipitate is filtered, and the excess of dichromate is determined by adding iodide and titrating the resulting iodine with standard thiosulfate. The derivation of this somewhat complicated calculation is as follows:

A measured portion of potassium dichromate solution is titrated with standard thiosulfate to give the "blank." The same volume of dichromate solution is added to the weighed sample, part of it is precipitated, and the excess is titrated to give the "titration." The number of equivalents of dichromate precipitated is given by the expression

$$\frac{(Blk\text{-}titr) \ (ml) \times N \ S_2O_3^{--} \ (eq/1)}{1000 \ (ml/1)}$$

This expression multiplied by the equivalent weight of dichromate gives the number of grams of dichromate precipitated. The figure obtained is then divided by the grams of dichromate precipitated by one mole of quaternary compound, shown by the equation given for the reaction to be one-half of the molecular weight of potassium dichromate (147.1). The result is the number of mols of quaternary salt in the sample. Multiplying this by the empirical molecular weight (400 for the triethyl compound) gives the grams of quaternary salt in the sample. The entire calculation is thus

$$\% \ Q = \frac{(Blk\text{-}titr)(ml) \times N \ S_2O_3^{--}(eq/1) \times (eq \ wt \ K_2Cr_2O_7)(g/eq) \times MWQ \ (g/mol) \times 100}{Sample \ Wt \ (g) \times 1000 \ (ml/1) \times g \ K_2Cr_2O_7/mol \ Q}$$

$$= \frac{(Blk\text{-}titr) \times N \ S_2O_3^{--} \times 49.03 \times 400 \times 100}{Sample \ Wt \times 1000 \times 147.1}$$

PROCEDURE

Weigh a suitable sample of the quaternary compound or solution, transfer it to a 100-ml volumetric flask, and dilute to the mark with water. With a pipette transfer a 20-ml aliquot to a 250-ml glass-stoppered Erlenmeyer flask. Transfer a 20-ml portion of standard 0.1 N $K_2Cr_2O_7$ to the flask, stopper it, and shake it for 5 minutes. Filter the precipitate on Whatman No. 41 paper, wash it thoroughly with water, receiving the filtrate and washings in a 500-ml Erlenmeyer flask. Dissolve 3 grams of KI in 50 ml of water, add to it 3 ml of 6N HCl, and immediately add this solution to the contents of the Erlenmeyer flask. Swirl, cover with a watch glass, and allow it to stand in the dark for 5 minutes. Dilute to about 400 ml, and titrate with standard $0.1N$ $Na_2S_2O_3$ to the end point for starch. To determine the volume of standard thiosulfate equivalent to the dichromate, make a blank titration with the same amount of dichromate used for the sample, and calculate the percentage of quaternary salt.

$$\% \, Q = \frac{(Blk\text{-}titr) \times N \, S_2O_3{}^{--} \times 1.666 \times MWQ}{Sample \, Wt \times 10}$$

NOTES: For solutions containing 30 to 60% of quaternary ammonium salts in the 300–400 molecular weight range a 1.5-gram sample is satisfactory.

The calculation takes account of the 20-ml aliquot; for the triethyl compound mentioned in the discussion the molecular weight (MWQ) would be 400.

References

1. J. S. BALL, M. L. WHISMAN, and W. J. WENGER, "Nitrogen Content of Crude Petroleums," *Ind. Eng. Chem.*, 43:2577 (1951).
2. L. H. DIMPFL, J. E. GOODRICH, and R. A. STAYNER, "Evaluating Additives for Distillate Fuel Oil—Storage Tests," *Ibid.*, 48:1885 (1956).
3. M. H. MUHS and F. T. WEISS, "Determination of Pyrrolic Nitrogen in Petroleum Distillates," *Anal. Chem.*, 30:259 (1958).
4. S. L. MEISEL, E. KOFT, JR., and F. G. CIAPETTA, "Effect of Nitrogen Compounds on Platinum-Acidic Oxide Catalysts," Preprints of General Papers, Division of Petroleum Chemistry, American Chemical Society, Vol. 2, No. 4, p. A-45, September 1957.
5. E. D. NOBLE, "Determination of Trace Kjeldahl Nitrogen in Petroleum Stocks," *Anal. Chem.*, 27:1413 (1955).
6. D. F. BOLTZ, Ed., *Colorimetric Determination of Nonmetals*, Interscience, New York (1958).
7. R. J. MILLER, "Hydrogenation of Petroleum Nitrogen Bases," *Ind. Eng. Chem.*, 43:1410 (1951).
8. J. L. DARRAGH and R. D. STAYNER, "Quaternary Ammonium Compounds from Dodecylbenzene," *Ind. Eng. Chem.*, 46:254 (1954).
9. P. L. KIRK, "Kjeldahl Method for Total Nitrogen," *Anal. Chem.*, 22:354 (1950).
10. G. R. LAKE, P. McCUTCHAN, R. VanMETER, and J. C. NEEL, "Effects of Digestion Temperature on Kjeldahl Analyses," *Ibid.*, 1634 (1951).
11. P. McCUTCHAN and W. F. ROTH, "Determination of Nitrogen," *Ibid.*, 24:369 (1952).
12. G. R. LAKE, "Determination of Nitrogen in Petroleum and Shale Oil," *Ibid.*, 24:1806 (1952).
13. H. WILFARTH, *Chem. Zentr.*, 56:17, 113 (1885).
14. J. W. GUNNING, *Z. anal. Chem.*, 28:188 (1899).
15. *Official Methods of Analysis of the Association of Official Agricultural Chemists*, 7th Ed., p. 13 (1950).

16. V. Z. DEAL, F. T. WEISS, and T. T. WHITE, "Determination of Basic Nitrogen in Oils," *Anal. Chem.*, 25:426 (1953).
17. C. W. PORTER, T. D. STEWART, and G. E. K. BRANCH, *The Methods of Organic Chemistry*, Ginn and Company, Boston (1927).
18. F. M. STUBBLEFIELD and E. E. DE TURK, "Effect of Ferric Sulfate in Shortening Kjeldahl Digestion," *Ind. Eng. Chem., Anal Ed.*, 12:396 (1940).
19. G. A. MILLS, E. R. BOEDEKER, and A. G. OBLAD, "Chemical Characterization of Catalysts. I. Poisoning of Cracking Catalysts by Nitrogen Compounds and Potassium Ion," *J. Am. Chem. Soc.*, 72:1554 (1950).
20. J. S. FRITZ, "Titration of Bases in Nonaqueous Solvents," *Anal. Chem.*, 22:1028 (1950).
21. D. B. BRUSS and G. E. A. WYLD, "Methyl Isobutyl Ketone as a Wide-Range Solvent for Titration of Acid Mixtures and Nitrogen Bases," *Ibid*, 29:232 (1957).
22. J. L. DREHER, B. W. HOTTEN, and C. F. CARTER, "A New Synthetic Thickener for Multipurpose Lubricating Grease," *Journal of the National Lubricating Grease Institute*, 20, No. 11:10, Feb., 1957.
23. O. B. HAGER, E. M. YOUNG, T. L. FLANAGAN, and H. B. WALKER, "Determination of High Molecular Weight Quaternary Ammonium Compounds as the Triiodides," *Anal Chem.*, 19:885 (1947).
24. E. FLOTOW, *Pharm. Zentralhalle*, 83:181 (1942).

Chapter 18

PHOSPHORUS

Phosphorus compounds are used as additives in a variety of petroleum products including greases, lubricating oils, and gasoline in a broad range of concentrations. Two special methods are of interest: Calkins and White [1] have described a spectrographic procedure for determining phosphorus in lubricating oils and additives; and Hoffman et al.[2] have given a method for determining phosphorus in gasolines containing tricresylphosphate as an inhibitor of surface ignition.

In Section 1 is discussed the decomposition of petroleum samples by ashing in the presence of zinc oxide as a capturing agent for phosphorus. It will presently be shown that this method, first described by Goodloe,[3] is effective for the majority of petroleum additives that contain phosphorus, and results compare favorably with the more generally applicable ASTM procedure.[4] As the peroxide bomb fusion method is not entirely suitable for petroleum products a detailed procedure is not given. Results of this method tend to be low as shown in Table 18:1.

As concentrations of phosphorus range from a few parts per million in gasoline to approximately ten per cent in some additives, the method of determination chosen must be appropriate for the amount of phosphorus to be handled. The heteropoly molybdenum blue method, suitable for the 0.001–0.05 per cent range is described in Section 2. The 0.05–2 per cent range is conveniently covered by the molybdivanadophosphoric acid method given in Section 3. Larger amounts are best handled by the familiar alkalimetric method discussed in Section 4.

Table 18:1 shows the results of four different methods of decomposition and two methods of determination applied to oil solutions of nine different phosphorus compounds. Comments relevant to these results will be found in pertinent sections of this chapter.

TABLE 18:1

COMPARISON OF PHOSPHORUS METHODS

Compound Phosphorus	Per Cent Phosphorus				
	Present	D1091 ASTM *	Volumetric Perox. Bomb	Volumetric ZnO Ashing	Color * ZnO Ashing
Tri-p-tolyl Phosphate	0.15	0.15	0.14	0.16, 0.17	0.15, 0.15
Unknown †	—	0.41	0.37	0.40	0.39
Calcium Cetyl-phosphate	0.79	0.78	0.72	0.76	0.78, 0.79
Tri-p-tolyl Phosphite	0.96	0.94	0.83	0.97, 0.97	0.97, 0.99
Triphenyl-phosphine	0.99	1.00	0.87	0.30, 0.37	0.34, 0.37
Zinc Dicetyl-phenyldithio-phosphate	1.05	1.03	1.02	1.01	1.06, 1.06
Zinc Dibutyl-dithio-phosphate	1.67	1.67	1.60	1.66	1.67, 1.70
Lecithin	—	—	2.06, 2.11	2.15, 2.18	2.10, 2.10
Zinc Dibutyl-dithio-phosphate			7.18, 7.18	7.29, 7.30	

* Molybdivanadophosphoric acid colorimetric procedure.
† This material also contained 3% chlorine and 10% sulfur.

Before proceeding with the details of the procedure, it will be convenient to provide first a table of sample sizes and appropriate procedures for the various concentrations of phosphorus normally encountered.

TABLE 18:2

SAMPLE SIZES FOR PHOSPHORUS DETERMINATION

% Phosphorus	Sample Size, Grams	Determination
0.001–0.005	1	Section 2
0.01 –0.05	0.1	Section 2
0.05 –0.1	2	Section 3
0.2 –2	0.2	Section 3
2 –8	0.2	Section 4

1. Decomposition by Ashing in the Presence of Zinc Oxide

The use of zinc oxide as a capturing agent for phosphorus was first applied to the determination of phosphorus in lubricating oils by Goodloe.[3] As originally proposed, the procedure consisted of weighing a sample in a porcelain crucible, covering it with an equal weight of zinc oxide, and burning the oil off slowly with a low flame. Phosphorus is retained as zinc pyrophosphate.

Calcium oxide is also an effective capturing agent, but it has two disadvantages which make its use undesirable: When determining phosphorus by the colorimetric procedures in Sections 2 and 3, wherein sulfuric acid is used, the relatively small solubility of calcium sulfate causes the latter to precipitate. When high-phosphorus additives are ashed in porcelain with lime, the crucible is attacked somewhat, introducing silica. The addition of nitric acid in the volumetric procedure in Section 4 precipitates silicic acid which clogs the filter paper, and any silica remaining in solution is subsequently precipitated as silicomolybdate, further slowing the filtration.

The efficiency of the ashing method with zinc oxide is shown in Table 18:1 wherein the results obtained with several different types of phosphorus compounds are tabulated. It will be noted that the procedure failed when applied to an oil solution of triphenylphosphine, but was satisfactory for tri-p-tolyl phosphite. Phosphine derivatives are not ordinarily used in petroleum products, however, and for the great majority of samples encountered, the procedure is applicable. Another exception that should be noted is cresyl diphenylphosphite; only about fifty per cent of the phosphorus is recovered by zinc oxide ashing.

Because of the necessity of producing suitable acidities for the colorimetric procedures in Sections 2 and 3 the zinc oxide must be weighed. The sample of oil is added to the zinc oxide in a porcelain crucible, and after the combustible material is burned off with a luminous flame the crucible is placed in a furnace and ignited at about 900°C for a few minutes. No phosphorus is lost even if the ignition is continued for many hours, but zinc oxide tends to sublime, and this contaminates the furnace if the ignition is unduly prolonged. When ashing a sample for the volumetric procedure in Section 4 the amount of zinc oxide is not critical; variations of two to six grams do not affect the result.

PROCEDURE

1. Transfer 1.8 gram of zinc oxide to a No. 0 Coors porcelain crucible. By means of a spatula transfer a suitable amount of sample (see Table 18:2) to the crucible, and tap gently until the sample is covered with zinc oxide. Place the crucible on a wire triangle and heat it gently with a luminous flame until burning ceases, then gradually increase the heat until a soft carbon ash remains. Transfer the crucible to a furnace and ignite for 5 minutes at about 900°C, then remove the crucible, and allow it to cool to room temperature.

> NOTES: For the colorimetric procedures in Sections 2 or 3 the zinc oxide should be weighed to the nearest 0.1 gram; for the volumetric procedure in Section 4, 2–5 grams may be used. If the latter procedure is to be applied, a No. 1 crucible is more convenient.
>
> If carbon remains the crucible should be ignited further until the residue is clean.

2. Proceed with the determination of phosphorus by the appropriate procedures in Sections 2, 3, or 4.

2. Heteropoly Molybdenum Blue Colorimetric Method

The method with heteropoly molybdenum blue for determining phosphorus has been the subject of many investigations. Boltz and Mellon [5] have made a study of the variables as applied to the determination of arsenic, germanium, phosphorus, and silicon. The procedure below follows essentially the recommendations of Fontaine.[6] The development of the blue color depends upon the catalytic effect of phosphate ion on the reduction of molybdate by stannous chloride. This reduction is slow at high acidity, but is hastened by heating, and the color produced is stable for 24 hours. The wavelength of maximum absorption depends upon the reducing agent used, and other factors; under the conditions prescribed here the maximum occurs at 820 mμ. The acidity may vary from 1.7 to 2.1 normal without altering the final color.

This procedure is recommended for phosphorus concentrations in the range of 0.001–0.05 per cent; if the approximate content is known, the optimum size of sample can be determined from the expression:

$$\text{Weight of Sample} = .004/\% \text{ P}$$

As moderate amounts of aluminum, barium, calcium, copper, iron, lead, molybdenum, and zinc are without effect, the procedure is applicable to both new and used lubricating oils. If barium, calcium, or lead occurs in larger amounts than are retained in solution their insoluble sulfates can be filtered and discarded without loss of phosphorus.

PREPARATION

1. REAGENTS:

   Ammonium Molybdate Solution: Dissolve 64 grams of $(NH_4)_6Mo_7O_{24} \cdot 4H_2O$ in a mixture of 300 ml of water and 500 ml of $10N$ H_2SO_4, and dilute to 1000 ml. This solution is stable at least 12 months.

   Stannous Chloride Solution: Dissolve 5 grams of $SnCl_2 \cdot 2H_2O$ in 25 ml of HCl. This solution should be renewed weekly.

   Stannous Chloride Solution, Dilute: Dilute 1.0 ml of the above solution to 100 ml with water. Prepare daily as required.

   Sulfuric Acid, 10N: Slowly add 280 ml of concentrated H_2SO_4 to 500 ml of water, mix, cool, and dilute to 1000 ml.

   Standard Phosphorus Solution: Dissolve 0.4392 gram of KH_2PO_4 in water, add a few small crystals of $KMnO_4$ as a preservative, and dilute to 1000 ml. 1 ml = 0.10 mg P.

2. STANDARDIZATION:

Dilute the prepared phosphorus standard tenfold to produce a solution containing 0.01 mg P/ml. Transfer .01, .03, .05, .07, .09 mg of phosphorus to a series of 100-ml beakers, and include a blank. To each beaker add 1.8 g of ZnO, and 20 ml of 10N H$_2$SO$_4$. Dilute to about 30 ml and boil gently for 2–3 minutes to dissolve the ZnO. Cool, and transfer to 100-ml volumetric flasks, diluting in the process to about 70 ml. To each flask add 10 ml of molybdate reagent, swirl to mix, and add 5 ml of dilute stannous chloride solution.

Suspend the flasks in a boiling water bath for 20 minutes, cool, and dilute to the mark with water. Measure the transmittances of the solutions at 820 mμ using 13-mm cuvettes, and plot the results against the corresponding mg of phosphorus on semi-logarithmic paper.

PROCEDURE

1. Transfer the contents of the crucible prepared by the procedure in Section 1 to a 100-ml beaker by tapping the bottom with the handle of a spatula. Add 10N H$_2$SO$_4$ to the crucible to dissolve any adhering ash, using a total of 20 ml of the acid, and warming, if necessary, to effect solution. Transfer the acid and crucible washings to the beaker, and boil it gently for 2–3 minutes to dissolve the ash completely. Cool, and transfer the solution to a 100 ml volumetric flask, filtering, if necessary, through Whatman No. 42 paper.

NOTES: In the presence of barium or lead, an insoluble residue will remain.

If barium sulfate is present it is finely divided, and requires a close-textured paper.

2. Add 10 ml of molybdate reagent, mix, and add 5 ml of dilute stannous chloride solution. Heat the flask for 20 minutes in a boiling water bath, cool, dilute to the mark, and measure the transmittance of the solution at 820 mμ using a 13-millimeter cuvette. From the prepared calibration curve read the corresponding milligrams of phosphorus.

NOTE: The color after heating is stable for at least 24 hours. Some spectrophotometers require a special filter to remove scattered light at the longer wave lengths.

3. Molybdivanadophosphoric Acid Colorimetric Method

The yellow color formed when ammonium molybdate and ammonium metavanadate are added to an acid solution of phosphate has been attributed to the formation of $(NH_4)_3PO_4 \cdot 16MoO_3$, but the exact formulation of the compound is uncertain. A critical study of the procedure has been reported by Kitson and Mellon.[7] The method is relatively free from error, and the color is stable for at least two weeks.

The procedure given here is appropriate for phosphorus contents in the range of 0.05–2 per cent. If the approximate concentration of phosphorus is known the optimum size of sample can be calculated from the expression:

Weight of Sample $= 0.15/\%P$

None of the elements likely to be present in new or used lubricating oils, or additives, interferes; barium and lead separate as insoluble sulfates, however, and must be filtered out if present.

PREPARATION

1. REAGENTS:

   Ammonium Metavanadate Solution: Dissolve 2.5 grams of NH_4VO_3 in 500 ml of boiling water, cool, add 20 ml of HNO_3, and dilute to 1000 ml with water.

   Ammonium Molybdate Solution: Dissolve 50 grams of $(NH_4)_6Mo_7O_{24} \cdot 4H_2O$ in 1000 ml of water.

   Sulfuric Acid, 10N: Slowly add 280 ml of concentrated H_2SO_4 to 500 ml of water, mix, cool, and dilute to 1000 ml.

   Standard Phosphorus Solution: Dissolve 4.392 grams of KH_2PO_4 in water, add a few small crystals of $KMnO_4$ as a preservative, and dilute to 1000 ml. 1 ml $= 1.0$ mg P.

2. STANDARDIZATION:

Transfer 0.5, 1.0, 2.0, 3.0 mg of phosphorus to a series of 100-ml beakers, and include a blank. To each beaker add 1.8

gram of ZnO, and 10 ml of $10N$ H_2SO_4. Dilute to about 40 ml, and boil gently for 2–3 minutes to dissolve the ZnO. Cool this solution and transfer it to 100-ml volumetric flasks, diluting in the process to about 70 ml. To each flask add 10 ml of NH_4VO_3 solution and 10 ml of molybdate solution, mixing after each addition, and dilute to volume.

Allow the solutions to stand about 30 minutes, and measure the transmittances at 470 mμ using 13-mm cuvettes. Plot the transmittances against the corresponding mg of phosphorus on semi-logarithmic paper.

PROCEDURE

1. Transfer the contents of the crucible prepared by the procedure in Section 1 to a 100-ml beaker by tapping the bottom with the handle of a spatula. Add $10N$ H_2SO_4 to the crucible to dissolve any adhering ash, using a total of 10 ml of the acid, and warming if necessary to effect solution. Transfer the acid and crucible washings to the beaker, and boil gently for 2–3 minutes to dissolve the ash. Cool, and transfer the solution to a 100-ml volumetric flask, filtering if necessary through Whatman No. 42 paper.

> NOTES: If barium or lead is present an insoluble residue will remain.
> A close-textured paper is required to retain barium sulfate.

2. Add 10 ml of NH_4VO_3 solution and 10 ml of molybdate reagent, mixing after each addition, then dilute to volume and let stand about 30 minutes. Measure the transmittance of the solution at 470 mμ, and read the corresponding mg of phosphorus from the prepared calibration curve.

4. Alkalimetric Determination of Phosphorus

The precipitation of ammonium phosphomolybdate followed by the volumetric determination of phosphorus is so well known that only a brief discussion is given here. The precipitate is formed by the addition of ammonium molybdate to a solution containing five to ten per cent of nitric acid and ammonium nitrate at a temperature between 20 and 45°C. The ideal composition of the precipitate is $(NH_4)_3PO_4{\cdot}12MoO_3$, and after filtering and washing it is dissolved in a measured excess of standard alkali with which it reacts as follows:

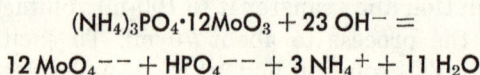

$$(NH_4)_3PO_4 \cdot 12MoO_3 + 23\,OH^- =$$

$$12\,MoO_4^{--} + HPO_4^{--} + 3\,NH_4^+ + 11\,H_2O$$

The excess of alkali is then titrated with standard acid using phenolphthalein indicator.

Most of the errors in this procedure cause high results; these include precipitation at too high a temperature, incomplete washing, loss of ammonia, and absorption of atmospheric carbon dioxide by the alkaline solution before titration. For a more comprehensive discussion of the method reference should be made to standard works on inorganic analysis.[8]

The procedure that follows is most convenient for materials that contain relatively high concentrations of phosphorus, such as lubricating oil additives, but it is also satisfactory for finished oils.

PREPARATION

1. REAGENTS:

Ammonium Nitrate Solution: Dissolve 800 grams of NH_4NO_3 in sufficient water to make one liter.

Molybdate Reagent: Add 265 ml of water and 50 ml of NH_4OH to a beaker, and sift 67 grams of MoO_3 into the solution while stirring. Prepare a second solution by diluting 265 ml of HNO_3 with 400 ml of water. Place the acid solution in a cooling bath and slowly add the ammonium molybdate solution, stirring continuously. Upon standing a yellow precipitate of $H_2MoO_4 \cdot H_2O$ separates. Filter before using.

Potassium Nitrate Wash Solution: Dissolve 10 grams of KNO_3 in 1000 ml of water, and add about 2 ml of 1% Aerosol wetting agent.

PROCEDURE

1. Transfer the contents of the crucible prepared by the procedure in Section 1 to a 400-ml beaker by tapping the

bottom with the handle of a spatula. Add 50 ml of water and a few drops of $N/10$ $KMnO_4$ to the beaker. To the crucible add 10 ml of HNO_3, and warm gently to dissolve any adhering ash, then transfer the acid to the beaker, rinsing the crucible with water. Add an additional 10-ml portion of HNO_3 to the beaker, mix thoroughly, and heat to boiling to dissolve all of the zinc oxide.

> NOTES: If the permanganate color disappears, add a few more drops until a permanent color is obtained.

2. Dilute the solution to about 200 ml, and add 20 ml of NH_4NO_3 solution. Adjust the temperature of the solution to about 40°C, and add 150 ml of filtered molybdate reagent. Stir the solution vigorously and allow to stand with occasional stirring until the precipitate settles rapidly.

> NOTES: The precipitation is hastened by heating, but the temperature must not exceed 45°C.
> With a clean solution precipitation is usually complete in 15–30 minutes. If the sample has been decomposed by peroxide fusion, however, several hours of standing may be required because of the retarding effect of the high concentration of salts.

3. Filter the precipitate on Whatman No. 7 filter paper, and wash the beaker and precipitate with eight or ten 10-ml portions of 1% KNO_3 wash solution. Transfer the paper and precipitate to the original beaker, add 100 ml of water, and a measured excess of standard 0.5 N NaOH. Stir to disintegrate the paper and dissolve the precipitate, add a few drops of phenolphthalein indicator, and titrate the excess alkali with standard 0.5 N HCl. Record the volumes and normalities of the two solutions, and calculate the percentage of phosphorus in the sample.

> NOTES: The precipitate has a pronounced tendency to creep; this is minimized by the Aerosol added to the wash solution.
> The equivalent weight of phosphorus is 1.347.

References

1. L. E. CALKINS and M. M. WHITE, "Analyze Additive Lubricants in Minutes Instead of Hours with Spectrographic Method," *Nat'l. Petroleum News*, 38, No. 27:519 (1946).
2. F. F. HOFFMAN, L. C. JONES, JR., O. E. ROBBINS, JR., and F. R.

212   THE INORGANIC ANALYSIS OF PETROLEUM

ALSBERG, "Colorimetric Determination of Phosphorous in Gasolines Containing Tritolyl Phosphate," *Anal. Chem.*, 30:1334 (1958).
3. P. GOODLOE, "Photometric Determination of Added Phosphorus in Oils," *Ind. Eng. Chem., Anal. Ed.*, 9:527 (1937).
4. AM. SOC. TESTING MATERIALS, *1955 Book of ASTM Standards*, Part 5, D1091–54T.
5. D. F. BOLTZ with M. G. MELLON, "Determination of Phosphorus, Germanium, Silicon, and Arsenic by the Heteropoly Blue Method," *Anal. Chem.*, 19:873 (1947).
6. T. D. FONTAINE, "Spectrophotometric Determination of Phosphorus," *Ind. Eng. Chem., Anal. Ed.*, 14:77 (1942).
7. R. E. KITSON with M. G. MELLON, "Colorimetric Determination of Phosphorus as Molybdivanadophosphoric Acid," *Ibid.*, 16:379 (1944).
8. W. F. HILLEBRAND, G. E. F. LUNDELL, H. A. BRIGHT, and J. I. HOFFMAN, *Applied Inorganic Analysis*, 2nd Ed., John Wiley and Sons, New York (1953).

Chapter 19

SELENIUM

As the presence of selenium has not been reported in crude oils,[1] occasions for determining the element in the petroleum laboratory are relatively infrequent. Dialkyl selenides, however, are effective inhibitors of oxidation in lubricating oil, and the analysis of the selenides themselves, or their oil blends is occasionally required. Denison and Condit [2] have investigated the mechanism whereby these agents inhibit oxidation in lubricating oil, and have briefly described the properties of several dialkyl selenides. These compounds have a high antioxidant activity with but little tendency toward the formation of engine deposits.

This chapter covers the determination of relatively small amounts of selenium in lubricating oils, and the analysis of dialkyl selenides. For small concentrations, a colorimetric procedure using phenylhydrazine is given in Section 1; and in Section 2 an iodometric method is provided for larger amounts of the element.

1. Colorimetric Determination of Selenium

Although the phenylhydrazine method for selenium is not particularly sensitive, it is easily followed, and it gives reproducible results when performed as prescribed below. The basis of the method is the production of a yellow sol of elemental selenium formed by reduction with phenylhydrazine, and stabilized by gum arabic. The development of the color requires about 30 minutes, thereafter increasing slightly during an additional hour of standing. The sample is decomposed by oxidation in the Parr oxygen bomb.

PREPARATION

1. REAGENTS:

> *Gum Arabic Solution:* Dissolve 5 grams of gum arabic in 100 ml of water. Filter if necessary.
>
> *Phenylhydrazine Hydrochloride Reagent:* Dissolve 5 grams of phenylhydrazine hydrochloride in 100 ml of water. The solution is unstable and should be discarded when it darkens.
>
> *Sodium Hydroxide 0.5N:* Dissolve 20 grams of NaOH in 1000 ml of water.
>
> *Standard Selenium Solution:* Dissolve 1.405 grams of SeO_2 in 1000 ml of water. 1 ml = 1.0 mg Se.

2. STANDARDIZATION:

Transfer 1.0, 2.0, 3.0, 4.0, and 5.0 mg of selenium to a series of 100-ml volumetric flasks, each containing 5 ml of $0.5N$ NaOH, and include a blank. Dilute to about 70 ml, add 3 ml of H_2SO_4, mix, and cool. To each flask add 2 ml of 5% gum arabic solution and 5 ml of 5% phenylhydrazine hydrochloride. Dilute to volume, mix, let stand 30 minutes, and measure the transmittances of the solutions at 425 mμ using 13-mm cuvettes. Plot the transmittances against the milligrams of selenium on semi-logarithmic paper.

PROCEDURE

1. Transfer a 1-gram sample to a fused silica combustion cup, and ignite in an oxygen bomb following the usual procedure,[3] except that 5 ml of $0.5N$ NaOH and 10 ml of water should be used to wet the inside of the bomb.

2. After ignition vent the excess oxygen carefully, and wash the contents of the bomb into a 100-ml volumetric flask (Note) with hot water. Add 3 ml of H_2SO_4, mix, and cool. Add 2 ml of 5% gum arabic solution, 5 ml of 5% phenylhydrazine reagent, dilute to volume, and mix. Let stand 30 minutes, and measure the transmittance of the solution at 425 mμ using a 13-mm cuvette. From the prepared calibration curve read the corresponding milligrams of selenium.

NOTE: If too much wash water is used the excess may be evaporated on a steam plate without loss of selenium.

2. Volumetric Determination of Selenium

The procedure that follows is intended for the determination of selenium in such compounds as the dialkyl selenides mentioned at the beginning of the chapter, in which the selenium content is relatively high (14–20 per cent). Organic material is destroyed by wet oxidation, but because of the volatility of selenium compounds the initial digestion must be carried out under a reflux condenser.

After the organic material has been destroyed the solution is diluted and potassium iodide is added. Selenious acid oxidizes an equivalent amount of iodide to iodine, and the latter is titrated potentiometrically with standard thiosulfate.

$$4\,H^+ + H_2SeO_3 + 6\,I^- = Se + 2\,I_3^- + 3\,H_2O$$

PROCEDURE

1. Transfer a suitable weighed sample (0.2–0.3 g) to a Kjeldahl flask, add a few glass beads, and 25 ml of HNO_3 Immerse the bulb of the flask in ice water, and then add slowly while swirling 10 ml of H_2SO_4. Place the flask on the electric heater, attach a West condenser, and reflux for about 30 minutes, then rinse and detach the condenser tube.

NOTE: The poisonous nature of selenium compounds should be taken into account in these operations. Use a hood with adequate draft, and avoid contact with the skin, or inhalation of vapors.

2. Boil off the HNO_3, heat to light fumes of H_2SO_4, and oxidize the organic material as usual with HNO_3 and 30% H_2O_2. When a clear solution is obtained, cool it, carefully add about 10 ml of H_2O, heat to light fumes of H_2SO_4, and cool to room temperature.

NOTE: Selenious acid is not oxidized to selenic acid by nitric acid, and is not volatilized from sulfuric acid even when heated to heavy fumes.

3. Transfer the cool acid solution to a 400-ml beaker containing 200 ml of water, and rinse the flask. Cool it, add 5 grams of urea, stir to dissolve this, then add 3 grams of KI.

Stir until the salt dissolves, cover with a watch glass, and let the solution stand 5 minutes in the dark.

4. Place the solution on a titration stand equipped with platinum and calomel electrodes and a mechanical stirrer. Titrate potentiometrically with standard 0.1 N $Na_2S_2O_3$ using the (plus) millivolt scale to the maximum change in potential. Calculate the selenium content of the sample, using an equivalent weight of 19.74.

NOTE: The maximum deflection usually occurs between 200 and 100 mv, and is of the order of 25 mv.

References

1. M. C. K. JONES and R. L. HARDY, "Petroleum Ash Components and Their Effect on Refractories," *Ind. Eng. Chem.*, 44:2615 (1952).
2. G. H. DENISON and P. C. CONDIT, "Oxidation of Lubricating Oils," *Ibid.*, 41:944 (1949).
3. AM. SOC. TESTING MATERIALS, *1955 Book of ASTM Standards*, Part 5, D129.

Chapter 20

SILICON

Silicon occurs in practically all crude oils [1] in both soluble and insoluble forms. The latter are often introduced from drilling muds, but sand and dirt are usually present also. Chemical methods are not convenient for the determination of small concentrations of silicon, mainly because of difficulties with the reagent blank, and for this reason a spectrochemical method, such as that described by Gunn and Powers,[2] is best applied to samples of crude oil, or feed stock for catalytic cracking.

In Section 1 the determination of silicon in silicate esters, silicones, and siloxanes is described, and in Section 2 the determination of silica in used lubricating oils is considered.

1. Determination of Silicon in Synthetic Oils

The use of organosilicon compounds as high-temperature hydraulic fluids for aircraft has been mentioned by Farrington.[3] Silicate esters, silicones, and siloxanes have more favorable characteristics for this service than petroleum oils. For example, their viscosity is low at low temperatures, their volatility is low at high temperatures, and additives are available for increasing their resistance to hydrolysis.

Rochow [4] has given a brief survey of the analytical methods for the silicones. Direct ashing is not satisfactory because the more volatile silicones are evolved as stable vapors, and the residue usually contains silicon carbide. The synthetic oils under consideration here are easily decomposed by a mixture of nitric and sulfuric acids, and the resulting silica is then determined by the usual gravimetric procedure. As some of these compounds tend to creep the oxidation is best carried out in an Erlenmeyer flask to minimize losses.

PROCEDURE

1. Weigh a suitable sample, and transfer it to a 250-ml Erlenmeyer flask. Add 10 ml of H_2SO_4 and 10 ml of HNO_3, swirl to mix, heat this on the steam plate until any initial reaction subsides, then transfer it to a hot plate and concentrate to fumes of H_2SO_4. Destroy any residual organic material by adding small portions of HNO_3, then cool. Wash down the sides of the flask with a small amount of water, and again heat to fumes.

NOTE: A 0.5-gram sample is usually satisfactory for the synthetic oils under consideration.

2. Remove the flask from the heat, cool it to room temperature, and carefully add 50 ml of water. To the solution add a drop or two of 1% laboratory Aerosol solution, and filter on Whatman No. 40 paper, washing with water that contains a few drops of the wetting agent.

NOTE: Unless a wetting agent is used it is extremely difficult to effect a quantitative transfer of the silicic acid.

3. Transfer the paper and precipitate to a tared platinum crucible, smoke off the paper in a radiator, and ignite the crucible in a furnace at 1100°C for 30 minutes. Cool it in a desiccator, and reweigh.

To the residue add 1–2 ml of 1:1 H_2SO_4, and 5 ml of 48% HF. Evaporate the excess acid, first on the steam plate, then on a hot plate, and finally ignite at 1100°C for 30 minutes. Cool in a desiccator, and reweigh, recording the loss in weight as SiO_2.

NOTES: Enough dilute acid should be added to cover the residue. The use of dilute acid moderates the effervescence, and the heat of dilution is much diminished when the hydrofluoric acid is added.

If the percentage of silicon is desired the gravimetric factor is 0.4672.

2. Determination of Silica in Used Lubricating Oils

As the major portion of any silicon in crankcase oil is usually in an insoluble form, physical examinations such as "dirt counts," or "particle size distribution" are ordinarily more informative than "chemical analysis." Considering the variety of combinations of elements that may be encountered in used lubricating oils a generally applicable procedure would necessarily involve a number of steps to achieve precise results. No attempt will be made to provide such a method, but instead a procedure is given for estimating the concentration of silica with an accuracy of about ten per cent.

In order to fix any volatile silicones, or the like, the homogenized sample is pretreated with a few drops of fuming sulfuric acid, and the organic material is destroyed by soft ashing and wet oxidation. The acid solution is diluted, and insoluble matter—which may include barium, calcium, and lead sulfates, as well as silicic acid—is filtered out. The residue is then extracted with hot ammonium acetate solution which, according to Hillebrand et al.,[5] provides a satisfactory separation of lead sulfate from silica.

The extracted residue is next ignited and weighed, after which the usual treatment with hydrofluoric acid is applied, and the ignited residue is then reweighed. Large amounts of barium sulfate tend to produce low results; otherwise results by this procedure are quite satisfactory.

PROCEDURE

1. Transfer a 20–30-gram portion of the well-mixed sample to a 400-ml beaker. Add 5 drops of fuming H_2SO_4, and warm the mixture on the steam plate for 10 minutes with occasional stirring. Remove the stirring rod, wipe it with a piece of filter paper, and drop the paper into the beaker. Transfer the beaker to a muffle can, and ignite the oil by heating the can with a flame. When the sample is consumed burn the loose carbon from the upper walls of the beaker, avoiding direct heating of the bottom or lower portion. Cool, add 10 ml of H_2SO_4, and fume strongly for 10 minutes, then

oxidize with HNO_3 and H_2O_2 as usual. The final volume of H_2SO_4 should be about 5 ml.

2. Cool the acid solution, dilute it to about 50 ml, and boil for 5 minutes. Cool, and filter on Whatman No. 42 paper, washing the precipitate with water, and discarding the filtrate.

> NOTE: The solution is boiled to ensure complete solution of iron and aluminum sulfates, and to overcome the tendency of both barium and lead sulfates to form supersaturated solutions.

3. Leach the precipitate with 50 ml of hot 50% NH_4Ac, and finally wash with hot water. Transfer the paper and precipitate to a tared platinum crucible, smoke off the paper in a radiator, and ignite it for 30 minutes at 1100°C. Cool and reweigh the crucible.

4. To the residue add 1–2 ml of 1:1 H_2SO_4, and 5 ml of 48% HF. Evaporate the excess acid, first on the steam plate, then on a hot plate, finally igniting at 1100°C for 30 minutes. Cool in a desiccator and reweigh, recording the loss in weight as SiO_2.

> NOTES: Enough dilute acid should be used to cover the residue. If the residue after treatment with hydrofluoric acid exceeds 100 mg the results for silica may be somewhat low.

References

1. M. C. K. JONES and R. L. HARDY, "Petroleum Ash Components and Their Effect on Refractories," *Ind. Eng. Chem.*, 44:2615 (1952).
2. E. L. GUNN and J. M. POWERS, "Ash Residues from Petroleum Catalytic Cracking Feed Stocks," *Anal. Chem.*, 24:742 (1952).
3. B. B. FARRINGTON, "Recent Advances in the Lubrication Field," *Lubrication Engineering*, Jan.–Feb. (1955).
4. E. G. ROCHOW, *Chemistry of the Silicones*, John Wiley and Sons, New York (1946).
5. W. F. HILLEBRAND, G. E. F. LUNDELL, H. A. BRIGHT, and J. I. HOFFMAN, *Applied Inorganic Analysis*, 2nd Edition, John Wiley and Sons, New York (1953).

Chapter 21

SODIUM

Most crude oils contain sodium, often present in emulsified brine, and also introduced by bentonites used as drilling muds. When present as the chloride it is objectionable because of the corrosive effect (see Chapter 11, Section 3c), and it is usually removed by a preliminary desalting treatment. Sodium in residual fuel oils increases the destructive effect of vanadium on fire brick. The element is a permanent poison for cracking catalysts, participating in an acid-base reaction with the active acidic centers of these catalysts; its normal concentration in heavy gas oils is in the range of 0.05–1 ppm. Sodium is one of several metals that have a deleterious effect on silver bearings and wrist-pin bushings in diesel locomotives. For this reason, special low-sodium lubricating oils are required for this service, and their sodium content is carefully controlled.

Although standard methods [1,2] are available for determining sodium in the products mentioned, certain modifications are necessary to broaden their scope and include a few special cases. In Section 1, the decomposition of the sample is covered; in Section 2, the gravimetric method employing the triple acetate salt is given; in Section 3, a general discussion of a simplified flame photometric procedure is provided.

1. Decomposition of the Sample by Direct Ashing

The only satisfactory method of decomposing petroleum samples for the determination of sodium is direct ashing in platinum. Lykken et al.[3] report that high results may be obtained when oils with high phosphorus contents are burned

221

in porcelain. Wet oxidation is entirely unsatisfactory as much sodium is always introduced from attack on the glassware.

Karchmer and Gunn [4] have investigated the volatility of sodium chloride and sodium sulfate at various temperatures. They found that sodium chloride was practically all volatilized in 30 minutes at 1800°F; about four per cent was lost in four hours at 1150°F; and about eight per cent, in twenty-four hours at 1000°F. The loss of sodium sulfate was negligible in twenty hours at 1300°F, but about 2.4 per cent was lost in twenty hours at 1800°F. Thus it is seen that with proper temperature control sodium can be recovered virtually completely by a direct ashing procedure.

The oil sample is weighed and transferred to a platinum dish of suitable size. The dish is placed on a quartz triangle, and the oil is ignited with a flame. After the fire burns out the loose carbon is burned by using the radiant heat of a blast burner rather than by direct application of the flame, in order to avoid overheating the ash. When the excess carbon has been burned from the sides of the dish the latter is transferred to a furnace maintained at a temperature of about 550°C to complete the ignition of carbonaceous material.

PROCEDURE

1. Transfer a suitable sample to a clean 100-ml platinum dish, place the dish on a quartz triangle, and ignite the oil with a flame. When the fire has burned out place a blast burner near the dish so that the soft carbon is burned from the sides by the radiant heat of the flame.

> NOTES: For finished oils a 50-gram sample is usually satisfactory for either gravimetric or flame photometric determination. Thus, if an oil contains 50 ppm of sodium, a 50-gram sample would yield 2.5 mg of sodium, equivalent to 167 mg of sodium zinc uranyl acetate. For additives a 1–2-gram sample is ordinarily appropriate.
>
> The carbon should be burned at a low temperature to prevent the volatilization of sodium salts, and to minimize attack on the platinum dish by other elements that may be present.

2. Transfer the dish to a furnace, and ignite it for 30–60 minutes at 550°C. Remove the dish from the furnace, cool

it, add a few drops of HCl and 20 ml of water, and boil gently for a few minutes. Transfer the solution to a 150-ml beaker, filtering if necessary, and continue by the procedure in Section 2.

> NOTES: Small amounts of carbon left in the residue are of no importance. It seldom does any good to ignite longer than 60 minutes, and sodium may be lost in part if the ignition is prolonged.
>
> The ash usually dissolves completely, but occasionally there is a residue. If the procedure in Section 2 is to be followed the solution need not be filtered; if a flame photometer is to be used any insoluble material should be filtered and discarded to avoid plugging the burner (see Section 3).

2. Gravimetric Determination of Sodium

The triple acetate method of Barber and Kolthoff [5] is particularly suitable for the determination of sodium in lubricating oils as it is free from interference by several elements likely to be present in these products, such as barium, calcium, sulfur, and zinc. The preliminary treatment of the solution with an excess of zinc carbonate, as recommended by Overton and Garrett,[6] eliminates aluminum, iron, lead, and phosphate. By adding citric or tartaric acid to the prepared solution just before precipitation, as prescribed by Hale,[7] the interference of molybdenum is avoided. As molybdenum additives are often used in "tune-up" oils, the presence of it is always a possibility in used crankcase oils.

PREPARATION

1. REAGENTS:

> *Zinc Uranyl Acetate Solution:* Dissolve 100 grams of $UO_2(C_2H_3O_2)_2 \cdot 2H_2O$ and 278 grams of $Zn(C_2H_3O_2)_2 \cdot 2H_2O$ in 27 ml of HAc and 900 ml of water. Warm to dissolve, cool solution, let it stand 24 hours, and filter as needed.
>
> *Alcohol Wash Solution:* Saturate Formula 30 alcohol with $NaZn(UO_2)_3(C_2H_3O_2)_9 \cdot 6H_2O$, and filter as needed.
>
> *Washed Zinc Carbonate:* Extract solid $ZnCO_3$ (reagent grade) with water in a Soxhlet extractor

for several days. Dry it in an oven, and store in a glass-stoppered jar.

PROCEDURE

1. Dilute the solution prepared by the procedure in Section 1 to about 50 ml, add a few drops of bromcresol green indicator, and neutralize with dilute NH_4OH until the indicator turns blue (pH about 5.4). Add 5 grams of washed $ZnCO_3$, mix thoroughly, and let it stand several hours, or overnight.

> NOTE: The zinc carbonate blank should be less than 10 mg; once the value has been determined it may be used until a new batch of zinc carbonate is prepared.

2. Filter the solution through Whatman No. 40 filter paper, washing with five or six small portions of water, receiving the filtrate in a 100-ml beaker. Evaporate the solution until crystallization occurs, then add 1 drop of HNO_3, and about 0.2 ml of 50% citric or tartaric acid.

> NOTES: If too much hydrochloric acid was used to dissolve the ash in Section 1, an excessive amount of ammonium salt will be present. This may be destroyed by adding nitric and hydrochloric acids, and evaporating to dryness.
> Nitric acid is added to ensure the hexavalency of molybdenum.

3. Add to the above solution ten times its volume of filtered zinc uranyl acetate solution stirring continuously. Allow this to stand for 30 minutes, or until the precipitate is granular and settles rapidly after stirring.

4. Filter the precipitate on a tared medium-porosity filter crucible, and wash with several small portions of the alcohol wash solution. Finally wash with 2–3 small portions of ethyl ether, wipe the crucible with a damp cloth, and place it in a desiccator for about 10 minutes. Reweigh the crucible, and calculate the percentage of sodium in the sample (gravimetric factor, 0.01495).

3. Determination of Sodium by Flame Photometer

A standard method [2] is available for determining sodium in residual fuel oil, which describes the use of two commer-

cial flame photometers: the Beckman (Model 10,300 or
9,200), and the Perkin-Elmer (Model 52A).

The Coleman Flame Photometer, Model 21 (Coleman Instruments, Inc., Maywood, Illinois), is an inexpensive instrument which has proved reliable for routine determination of sodium in various products, including lubricating
oils, additives, and distillates. This is a filter photometer
with a simple atomizer-burner that provides a remarkably
stable flame. Natural gas and oxygen are required, as well
as an external galvanometer; internal standards are not
used.

To construct a standard curve, solutions are prepared
containing 0.5, 1.0, 1.5, and 2.0 mg of sodium per 100 ml.
Water is aspirated through the burner, and the galvanometer
is adjusted to zero using the controls on the photometer.
The standard solution containing 2.0 mg of sodium per 100
ml is aspirated through the burner, and the galvanometer
is set at an arbitrary luminosity reading of 40. The remaining standard solutions are then checked, and the corresponding luminosities are plotted on rectangular coordinate paper.

A sample prepared by the procedure in Section 1 is filtered into a 100-ml volumetric flask, and diluted to volume.
The galvanometer is adjusted as described, then the sample
solution is aspirated, and the luminosity reading recorded.
The corresponding number of milligrams of sodium in the
sample solution is then read from the prepared calibration
curve.

If potassium determinations are to be made a standard
solution containing both potassium and sodium may be prepared by dissolving 2.38 grams of KCl, and 0.508 gram of
NaCl in 1000 ml of water. A 10-ml aliquot of this solution,
diluted to 100 ml, gives a solution containing 12.5 mg of
potassium and 2.0 mg of sodium per 100 ml. An appropriate
luminosity setting for potassium is about 50; a potassium
filter is, of course, used when determining potassium.

No further details will be given here. Table 21:1 shows
comparative results by ASTM Method D1026, and by the
procedure outlined here.

TABLE 21:1

COMPARISON OF METHODS FOR SODIUM

Sample	ASTM D1026	Flame Photometer
1	0.0030	0.0029
2	0.0032	0.0032
3	0.0041	0.0036
4	0.0025	0.0024
5	0.0035	0.0037
6	0.0035	0.0030
7	0.0023	0.0019
8	0.0043	0.0035
9	0.106	0.103
10	0.090	0.088

References

1. AM SOC. TESTING MATERIALS, *1955 Book of ASTM Standards*, Part 5, D1026–51.
2. AM. SOC. TESTING MATERIALS, *Ibid.*, D1318–54T.
3. L. LYKKEN, K. R. FITZSIMMONS, S. A. TIBBETTS, and G. WYLD, "The Determination of Metals in Lubricating Oils," *Petroleum Refiner*, 24:405 (1945).
4. J. H. KARCHMER and E. L. GUNN, "Determination of Trace Metals in Petroleum Fractions," *Anal. Chem.*, 24:1733 (1952).
5. H. H. BARBER and I. M. KOLTHOFF, "A Specific Reagent for the Rapid Gravimetric Determination of Sodium," *J. Am. Chem. Soc.*, 50:1625 (1928).
6. O. R. OVERTON and O. F. GARRETT, "Determination of Sodium," *Ind. Eng. Chem., Anal. Ed.*, 9:72 (1937).
7. C. H. HALE, "Determination of Sodium in the Presence of Molybdenum," *Ibid.*, 15:516 (1943).

Chapter 22

SULFUR

Sulfur is an extremely important element in petroleum refining, and many methods have been proposed for determining it. No attempt will be made to review this field, but several of the more useful special methods will be mentioned as well as the well-established standard procedures. The great variety of sulfur compounds in petroleum is shown by an investigation by Thompson et al.,[1] who have separated and identified forty-three different compounds in a Texas crude oil.

The importance of sulfur arises mainly from its adverse effects in several different refining operations and products. Among these may be mentioned: 1) the corrosive action of hydrogen sulfide and other sulfur compounds; 2) induced formation of sludge in fuel oils; 3) inhibition of lead susceptibility by disulfides in gasoline; 4) adverse effect of disulfides on stability of color in gasolines; 5) objectionable odor of thiols; and 6) poisoning of catalysts by many sulfur compounds.

Sulfur additives are used for improving extreme pressure lubricating properties; they permit extremely heavy loads without seizure of lubricated surfaces. Other elements that may be present in these additives include barium, chlorine, lead, and phosphorus, among others.

By far the most important determination is that of total sulfur, and several standard procedures are available. The oxygen bomb method[2] is widely used for lubricating oils, additives, greases, crude oils, fuel oils, and other products. The original procedure specifies a gravimetric finish, but Kreider and Foulds[3] have described a volumetric procedure wherein the sulfate is reduced to sulfide by hydriodic acid and the evolved hydrogen sulfide is determined iodometrically. Phosphorus does not interfere, but barium, nitrogen, and selenium must be absent.

For the determination of sulfur in liquid products such as gasoline, gas oils, diesel fuels, etc., the lamp method [4] is usually applied. This procedure is not suitable for aromatic products, however, and for these the procedure described here in Section 2 should be used. The Esso lamp method [5] is an improved procedure which provides optional nephelometric, conductometric, and gravimetric finishes.

An especially rapid and accurate method for determining total sulfur employs the commercial Dietert apparatus, available from the H. W. Dietert Company, Detroit, Michigan. This method, first described by Erickson and Lindberg,[6] has been tentatively adopted by the ASTM.[7] The procedure is applicable to practically all petroleum products that can be weighed in an open boat, and in addition, if a fluxing agent is added, it ensures quantitative recovery of sulfur from inorganic salts such as barium sulfate. Under some conditions, excessive amounts of chlorine, or nitrogen may interfere.

Having mentioned several methods for determining total sulfur a few special procedures for determining different types of sulfur compounds will be noted. Ball [8] has developed a scheme for the analysis of sulfur groups, including hydrogen sulfide, thiols, elemental sulfur, aliphatic sulfides, and others. Bartlett and Skoog [9] have devised a very simple procedure for determining elemental sulfur in petroleum fractions. The concentration of mercaptans in streams of refinery gas is sometimes of interest; a convenient method for this determination has been given by Ellis and Barker.[10] The inhibiting effect of disulfides on the activity of tetraethyllead in gasoline has been mentioned; Earle [11] has given a method for the determination of disulfides in the presence of thiols. A great many more special methods have been proposed, but no more of them will be mentioned here.

For the determination of sulfur in additives that also contain high concentrations of chlorine the sample is best decomposed by peroxide bomb, as prescribed in the procedure in Section 1 below. As mentioned before, lamp methods are not suitable for determining sulfur in aromatic products, nor are they convenient for very small concentrations. To handle these materials the procedure in Section 2 is provided.

1. Determination of Sulfur by Peroxide Fusion Bomb

As materials that yield large amounts of chlorine attack the walls of oxygen bombs, they cannot be analyzed for sulfur by the Dietert method. Typical examples are additives for extreme pressure lubricants. These materials are best decomposed by peroxide fusion, and after separating heavy metals, such as barium or lead, the sulfur is determined by the precipitating and weighing of barium sulfate. If barium or lead is present it is metathesized and separated as the carbonate by boiling with ammonium carbonate, and filtering. As considerable carbonate is formed in the fusion most of the initial insoluble material will be in the form of carbonate, but some sulfate is lost if the boiling is omitted.

The use of picric acid to improve the precipitation of barium sulfate was first described by Lindsly.[12] Lincoln *et al.*[13] used picric acid after combustion in peroxide bomb to hasten precipitation, and minimize coprecipitation, but a recent study by Fischer and Rhinehammer [14] of the effect of picric acid and other addition agents on barium sulfate indicates that the usual errors of coprecipitation are not eliminated. The reagent does produce an easily filterable precipitate, however, and it is prescribed in the procedure given here.

SAFETY PRECAUTIONS

1. The eyes, face, and hands must be protected while preparing the fusion mixture.
2. The full 15 grams of sodium peroxide must always be used.
3. The sample must be dry and non-acidic, or else enclosed in a gelatine capsule.
4. Not more than 0.5 gram of total combustible material should be used.
5. Fusion cups should be discarded when their walls become eroded.
6. The gasket must be in good condition, and produce a tight seal between the fusion cup and the upper edge of the bell

body. Most explosions are caused by faulty gaskets.

7. The ignition switch should be located several feet from the bomb.

PREPARATION

1. APPARATUS:

> *Parr Peroxide Bomb* and accessories, 22-ml electric-
> ignition type
> *Ignition Transformer*, with ignition switch
> *Fuse Wire*
> *Water Bath*, circulating

2. FUSION MIXTURE FOR SULFUR:

Prepare this mixture immediately before using, by combining in a small glass-stoppered bottle, 15 grams of Na_2O_2, 1.0 gram of $KClO_4$, and 0.2 gram of powdered sugar, or benzoic acid. Mix rapidly and thoroughly with a stirring rod, and stopper to prevent absorption of moisture. Use as soon as possible.

PROCEDURE

1. Attach a 7-cm length of fuse wire to the terminals on the cover of the bomb so that the wire loop will extend a short distance into the fusion mixture when the cover is in place.

2. Weigh a maximum of 0.3 gram of the sample, and transfer it to a gelatine capsule (size 00). Transfer approximately half of the prepared fusion mixture to the dry fusion cup, and place the closed capsule upright in the fusion mixture, using tongs or forceps *(Not the fingers)*. Add the balance of the fusion mixture, and tap the cup lightly to pack the charge, then dust a *small amount* of sugar or benzoic acid on the surface.

> NOTES: Be sure that there is no peroxide clinging to the walls near the top of the cup; if any is in contact with the gasket the bomb may rupture upon ignition.
>
> A very small amount of combustion aid on the surface of the charge will ensure ignition.

3. Place the cover in position, adjusting the fuse wire so that it dips into the fusion mixture a short distance. See

that the cover is seated properly, and that the gasket seals both the cup and the upper edge of the bell body. Tighten the screw cap firmly with a wrench, and place the bomb in the water bath so that the ignition arm makes contact with the bomb head assembly. Ignite the charge with the ignition transformer, and allow the bomb to remain in the bath for about 10 minutes, or until cool. Remove the assembly from the bath, and wipe off the excess water.

4. Open the bomb, and if the fusion appears satisfactory, place the cup on its side in a 400-ml beaker, add 100 ml of water, and immediately place a cover glass on the beaker. When the initial vigorous reaction subsides, heat the beaker gently until the melt dissolves, then remove the cup, and wash it thoroughly with water. Dilute the solution to a volume of 150 ml, and boil this for about 15 minutes, or until excess peroxide has decomposed.

> NOTE: A few particles of carbon may be disregarded, but if any unburned material is visible the ignition must be repeated with another sample.

5. Cool the solution in an ice bath and neutralize to methyl orange with HCl, then add 15 grams of $(NH_4)_2CO_3$, and boil gently for 30 minutes. Filter the precipitate and wash with water containing a little Na_2CO_3. Neutralize the filtrate with HCl, add 5 ml in excess, and boil out the excess CO_2. Dilute the solution to about 300 ml, add 15 ml of a saturated aqueous solution of picric acid, heat to boiling, and precipitate $BaSO_4$ by adding drop by drop a 10% solution of $BaCl_2$. Continue boiling for about 5 minutes, then set the beaker aside for one hour.

> NOTES: This treatment is necessary only if barium or lead is present.
> A 10-ml portion of barium chloride solution is usually sufficient.

6. Filter the precipitate on Whatman No. 42 paper, washing with water. Transfer the paper and precipitate to a tared porcelain crucible, smoke off the paper in a radiator, then ignite the crucible in a furnace at 900°C for 30 minutes. Cool it, reweigh, and calculate the per cent of sulfur in the sample, correcting for a blank carried through the entire procedure (gravimetric factor for sulfur, 0.1373).

2. Determination of Sulfur by Wickbold Oxhydrogen Apparatus

The operation of the Wickbold combustion apparatus is mentioned in Chapter 2, Section 9, and described in detail in the Appendix. In this section the application of this method of combustion for the determination of sulfur in aromatic stocks, naphthas, kerosenes, and the like is described. This method is especially suited to the determination of low concentrations of sulfur as large samples can be rapidly burned. The colorimetric determination using barium chloranilate [15] gives satisfactory results, as can be seen in Table 22:1. This procedure is analogous to that used in Chapter 11, Section 6, for chloride. Barium is displaced from its chloranilate salt through the formation of the less soluble barium sulfate, at the same time releasing the red colored chloranilic acid anion at a concentration proportional to the amount of sulfate. The useful range is from one to seven milligrams of sulfur.

TABLE 22:1

COMPARISON OF METHODS FOR SULFUR

Sample	ASTM D1266	Wickbold
Light Gas Oil No. 1	0.23%	0.22%
Light Gas Oil No. 2	0.24%	0.24%
Alkane	0.010%	72 ppm
Kerosene No. 1	10 ppm	9 ppm
Kerosene No. 2	21 ppm	17 ppm
Kerosene No. 3	25, 29 ppm	23, 26 ppm
Alkylate	101 ppm	94, 100 ppm

PREPARATION

1. REAGENTS:

Ammonium Hydroxide, dilute: Add one volume of NH₄OH to ten volumes of water.

Barium Chloranilate (2,5-Dichloro-3,6-dihydroxy-p-benzoquinone Barium Salt): Eastman No. 7508.

2,4-Dinitrophenol Indicator: 0.1% aqueous.

Methyl Isobutyl Ketone: Shell Chemical Company.

Phenyl Sulfide: Eastman No. 619. Theoretical sulfur content 17.2%.

Potassium Acid Phthalate, 0.05M: Dissolve 10.2 grams of $KHC_8H_4O_8$ in 1000 ml of water.

Standard Sulfur Solution: Accurately weigh about 0.6 gram of phenyl sulfide, and transfer to a one-liter volumetric flask containing 500 ml of methyl isobutyl ketone, dilute to volume with the same solvent, and mix thoroughly.

mg S/ml = wt of phenyl sulfide × 0.172

2. STANDARDIZATION:

Charge the Wickbold absorber with 50 ml of water and 5 drops of 30% H_2O_2, and successively burn (for procedure, see the Appendix) the volumes of standard sulfur solution which will yield 1.0, 3.0, 5.0, and 7.0 mg of sulfur. Burn also 50 ml of methyl isobutyl ketone as a blank. Transfer the absorber solutions to Erlenmeyer flasks, and evaporate to a volume of about 25 ml. Cool, add several drops of 2,4-dinitrophenol indicator, and neutralize the solutions by adding drops of 1:10 NH_4OH until one drop turns the solution yellow. Transfer the neutralized solutions to 100-ml volumetric flasks, using a minimum of wash water.

Add 10 ml of 0.05*M*-potassium acid phthalate, 50 ml of Formula 30 alcohol, and dilute to the mark with water. Sift 0.2 gram of solid barium chloranilate into each flask, stopper it, and shake intermittently for 15 minutes. Filter the solutions through Whatman No. 42 paper, and read the transmittances at 530 mμ using 13-mm cuvettes. Plot the transmittances against the corresponding milligrams of sulfur on semi-logarithmic graph paper.

PROCEDURE

1. Weigh a suitable sample, and transfer it to an Erlenmeyer flask. Charge the Wickbold absorber with 50 ml of water and 5 drops of 30% H_2O_2, and burn the sample by the procedure described in the Appendix.

2. Drain the contents of the absorber into an Erlenmeyer

flask, rinsing the spray trap and absorber with water. Evaporate the solution and washings to a volume of about 25 ml, cool the solution, and proceed as directed further under *STANDARDIZATION*. Determine the number of milligrams of sulfur in the sample from the prepared calibration curve.

References

1. C. J. THOMPSON, H. J. COLEMAN, H. T. RALL, and H. M. SMITH, "Separation of Sulfur Compounds from Petroleum," *Anal. Chem.*, 27:175 (1955).
2. AM. SOC. TESTING MATERIALS, *1955 Book of ASTM Standards*, Part 5, D129–52.
3. R. E. KREIDER and J. G. FOULDS, "Bomb-Volumetric Method for Sulfur in Refined Petroleum Products," *Anal. Chem.*, 26:1983 (1954).
4. AM. SOC. TESTING MATERIALS, *ASTM Standards on Petroleum Products and Lubricants*, D1266–57T (1958).
5. C. C. HALE, E. R. QUIRAM, J. E. McDANIEL, and R. F. STRINGER, "Esso Lamp Method for Sulfur," *Anal. Chem.*, 29:383 (1957).
6. H. J. ERICKSON and R. E. LINDBERG, "Unit Cuts Sulfur Analysis Time by 94%," *Pet. Processing*, 9:1087 (1954).
7. AM. SOC. TESTING MATERIALS, *ASTM Standards on Petroleum Products and Lubricants*, D1552–58T (1958).
8. J. S. BALL, "Determination of Types of Sulfur Compounds in Petroleum Distillates," Bureau of Mines: Report of Investigations No. 3591 (1941).
9. J. K. BARTLETT and D. A. SKOOG, "Colorimetric Determination of Elemental Sulfur in Hydrocarbons," *Anal. Chem.*, 26:1008 (1954).
10. E. W. ELLIS and T. BARKER, "Determination of Thiols in Hydrocarbon Gases," *Ibid.*, 23:1777 (1951).
11. T. E. EARLE, "Determination of Disulfides in the Presence of Thiols," *Ibid.*, 25:769 (1953).
12. C. H. LINDSLY, "Determination of Sulfur in Rubber Compounds," *Ind. Eng. Chem., Anal. Ed.*, 8:176 (1936).
13. R. M. LINCOLN, A. S. CARNEY, and E. C. WAGNER, "Procedure for Semimicro determination of Sulfur in Organic Compounds," *Ibid.*, 13:358 (1941).
14. R. B. FISCHER and T. B. RHINEHAMMER, "Additive Agents in Analytical Precipitations," *Anal. Chem.*, 26:244 (1954).
15. J. E. BARNEY II and R. J. BERTOLACINI, "Ultraviolet Spectrophotometric Determination of Sulfate, Chloride, and Fluoride with Chloranilic Acid," *Ibid.*, 30:202 (1958).

Chapter 23

VANADIUM

Although vanadium is associated particularly with asphalt or asphalt-containing crudes, practically all crude oils contain some vanadium, its concentration ranging from a few tenths to several hundred parts per million. As much of it is non-volatile, it tends to concentrate in fuel oils, and other residua where it is undesirable because of its destructive effect on fire brick. Vanadium pentoxide is an effective flux, and its low melting point (690°C), combined with its tendency to form lower-melting eutectic mixtures with other materials, contributes to the formation of slag which raises problems when burning fuel oils that contain vanadium.

Studies by Beach and Shewmaker [1] have shown that vanadium is present in crude oils combined in porphyrins, some of which have low enough boiling points to permit the element to distill. Thus, it is found in gas oil cuts used as feed stock for catalytic cracking. In common with other elements of the first transition series, vanadium is a dehydrogenation catalyst with an activity from one-tenth to one-fourth that of nickel. Consequently it increases the formation of coke and hydrogen in cracking units with concomitant decrease of gasoline yield.

The determination of vanadium in distillates is covered in Section 1, and in Section 2 a procedure suitable for residual fuels and crude oils is provided. The great interest in vanadium problems has led to the preparation of synthetic oil-soluble vanadium compounds for purposes of research, and as these are occasionally submitted for analysis, a procedure suitable for higher concentrations of the element is included in Section 3.

1. Determination of Vanadium in Distillates

The existence in heavy distillates of volatile vanadium porphyrins has been established, and as a consequence direct ashing is not a satisfactory method for decomposing these samples. In Chapter 2, Table 2:4 shows the effect of pretreatment with sulfuric acid on the recovery of vanadium from gas oils. Results after pretreatment of four samples average seventy per cent higher than those obtained by free burning followed by wet oxidation. Wet ashing presumably gives complete recovery, but it is extremely tedious; Table 23:1 (see Section 2) shows that where sulfuric acid pretreatment is applied to fuel oils it gives comparable results.

The concentration of vanadium in heavy gas oils usually lies in the .05–.5 ppm range, and to avoid excessively large samples a sensitive colorimetric method is desirable. Wrightson[2] used diphenylbenzidine to determine vanadium in distillates, but 3,3′-dimethylnaphthidine is a superior reagent in some respects. Feigl[3] has applied its reaction with vanadium (V) to produce the blue-violet p-quinoidal diimine as a spot test for vanadium.

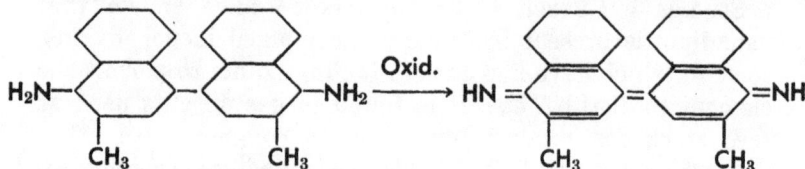

It is important that other oxidizing agents, including iron (III), which produce the same reaction, must be absent. As the colorimetric procedure given here has not been previously published, some of the details are next considered.

At the outset an attempt was made to use perchloric acid for oxidizing vanadium, which is satisfactory for amounts of the order of 1–3 mg as described in Section 2. Reduction products of perchloric acid, however, include hydrochloric acid,[4] and hydrogen peroxide,[5] both of which reduce vanadium (V). In the .01–.06 mg range under consideration here erratic results are obtained because with these small amounts of vanadium even this relatively slight reduction is significant. After considerable experimenting permanganate was chosen as the oxidant because the excess can be easily reduced with a drop of alcohol; the sensitivity of the reagent makes it essential that no oxidizing agent other than vanadate be present.

Iron is extracted as the thiocyanate by shaking with n-butanol, after which the aqueous layer is evaporated to destroy the excess thiocyanate. After oxidizing with permanganate and reducing the excess with alcohol, the solution is transferred to a small volumetric flask, and the reagent is added. The color requires about one hour to develop, and it is then stable for several hours. Because of the sensitiveness of the reagent and the rather critical oxidizing conditions, if the color obtained is too dark, or very light, it is advisable to transfer the contents of the flask back to a beaker, evaporate to fumes, and repeat the entire procedure.

Vanadium standard solutions are usually prepared from vanadium pentoxide. Vanadyl sulfate solutions seem to show greater stability, however, and the use of this salt is recommended here. Whichever is used, the prepared solution should be standardized by a volumetric method.

PREPARATION

1. REAGENTS:

   3,3'-Dimethylnaphthidine Reagent: Weigh 10 mg of 3,3'-dimethylnaphthidine, Eastman No. 7333, and transfer to a 100-ml mixing cylinder. Add 100 ml of 85% H_3PO_4, and mix. Allow this to stand overnight before using. The reagent darkens in color, but is usable for about one week.

   Potassium Thiocyanate Solution: Dissolve 100 grams of KSCN in 1000 ml of water.

   Vanadyl Sulfate: Amend Drug and Chemical Co., Inc., New York. Water contents of vanadyl sulfate vary; this product is approximately $VOSO_4 \cdot 4\frac{1}{2}H_2O$.

   Vanadium Standard Solution: Dissolve 4.8 grams of vanadyl sulfate in 200 ml of water containing 5 ml of H_2SO_4, warming to hasten solution. Cool, dilute to 1000 ml, and mix thoroughly. Using a 25-ml aliquot, standardize the solution by the volumetric procedure in Section 3.

$$mg \ V/ml = \frac{ml \ KMnO_4 \times N \times 16.98}{25.0}$$

2. STANDARDIZATION:

1. By suitable dilution of the vanadium standard solution prepared above, transfer .01, .02, .03, .04, and .05 mg of vanadium to 100-ml beakers, including a blank. Add 5 ml of H_2SO_4 and 2 drops of $HClO_4$, and evaporate to heavy fumes. Cool, add 10 ml of water and enough $N/10$ $KMnO_4$ to render the solution permanently pink. Allow the solution to stand 10 minutes, then add 1 drop of alcohol to reduce the excess permanganate, and let it stand for 10 minutes longer.

NOTES: As the permanganate oxidation of vanadium is slow the reaction should be allowed to proceed for the full ten minutes.
The solution should be allowed to stand for the full ten minutes extra to ensure complete reduction of permanganate. An amount of permanganate invisible to the eye produces an intense color with the reagent.

2. Transfer the solutions to 25-ml volumetric flasks, diluting in the process to about 23 ml. Mix, and add 1.0 ml of 3,3'-dimethylnaphthidine reagent. Dilute to volume, and allow the color to develop for one hour. Measure the transmittances of the solutions at 550 mμ, and plot the results against the corresponding milligrams of vanadium on semilogarithmic graph paper.

PROCEDURE

A. PRELIMINARY TREATMENT

1. Heat the sample to reduce its viscosity, and shake thoroughly. Transfer a 100-gram portion to a clean, dry, 400-ml beaker. Place the beaker on the steam plate, and add 10 drops of fuming H_2SO_4. Stir thoroughly, and then heat it for about 10 minutes with occasional stirring. Remove the

stirring rod, wipe with filter paper, and drop the paper into the beaker.

2. Place the beaker in a muffle can, and ignite the oil with a flame. As the sample burns down it may be necessary to apply the flame to maintain burning. Do this by placing the burner near the can, but do not heat the bottom directly. Remove the beaker from the can, and burn the carbon from the upper walls with a blast burner. Cool the beaker, add 15–20 ml of H_2SO_4, and allow this to digest on the hot plate for 15–20 minutes. Oxidize as usual with HNO_3, 30% H_2O_2, and finally with $HClO_4$. The final volume of H_2SO_4 should be about 5 ml.

3. Cool the acid solution, dilute to about 25 ml with water, and boil it gently for 5 minutes. Cool, filter if necessary, and transfer to a 125-ml separatory funnel, diluting in the process to about 50 ml.

B. REMOVAL OF IRON

Add enough $N/10$ $KMnO_4$ to the solution obtained in (A) to produce a permanent pink color, then add 2 ml of 10% KSCN solution, and mix. Add 25 ml of n-butanol, shake for 1 minute, and allow the layers to separate. Draw the aqueous layer into a 150-ml beaker, and evaporate to sulfuric acid fumes. If the solution darkens, add a small amount of 30% H_2O_2, fume again, and cool to room temperature.

NOTE: The excess of thiocyanate must be destroyed before proceeding. Use a hood for this operation.

C. DETERMINATION OF VANADIUM

1. To the cool H_2SO_4 add 10 ml of water, and enough $N/10$ $KMnO_4$ to produce a permanent pink color. Allow the solution to stand 10 minutes then add 1 drop of alcohol to reduce the excess permanganate, and let it stand an additional 10 minutes.

NOTE: A drop of the usual alcoholic solution of phenolphthalein indicator is a convenient reducing agent.

2. Transfer the solution to a 25-ml volumetric flask, diluting in the process to about 23 ml. Mix, add 1.0 ml of

3,3'-dimethylnaphthidine reagent, dilute to the mark, and allow the color to develop for one hour. Measure the transmittance of the solution at 550 mμ, and read the corresponding milligrams of vanadium from the prepared calibration curve.

2. Determination of Vanadium in Fuel Oils

The concentration of vanadium in fuel oils is of interest because of the destructive effect of its oxides and salts on fire brick,[6] and associated slagging problems in oil-fired marine boilers.[7,8] As the concentration of vanadium in residual fuel oils is dependent on the crude oil from which it is made, a wide range of concentrations is encountered.

It might be anticipated that the element would occur in a non-volatile form in residual products, but the data in Table 23:1 show that a small percentage is lost by free burning. The wet ashing procedure originally recommended by Milner et al.,[9] and subsequently modified by Gamble and Jones,[10] involves sludging the sample in a Vycor beaker with sulfuric acid, and finally igniting in a furnace at about 500°C until all carbon is burned. This procedure has been adopted as a tentative standard method,[11] with the phosphotungstate colorimetric method for the final determination. This method of decomposition is tedious for samples larger than about 25 grams, and the data in Table 23:1 show that pretreatment of the oil with sulfuric acid, followed by soft ashing and wet oxidation, gives comparable results.

TABLE 23:1

EFFECT OF METHOD OF DECOMPOSING SAMPLE
ON VANADIUM RECOVERED

Fuel Oil Sample	ppm Vanadium		
	Wet Ashing	Direct Ashing	H_2SO_4 Pretreat.
1	16	14	13
2	128	104	129
3	181	140	184
4	314	289	310
5	73	70	73
6	139	130	141

Wright and Mellon [12] have made an extensive study of the phosphotungstate method for vanadium, and although it is not so sensitive as that described in Section 1, it gives very reliable results when handling larger amounts of vanadium (1–2.5 mg). The method is relatively free from interferences; none of the elements normally found in fuel oil cause any difficulty. Excessive amounts of iron cause high results; potassium ion precipitates phosphotungstate.

To produce a color with phosphotungstate, vanadium must be pentavalent, and it is important to realize that vanadium (V) is reduced by fuming sulfuric acid. Table 23:2 shows the results of several oxidation processes as applied to 1.00 mg of vanadium; it is seen that perchloric acid is the most satisfactory oxidizing agent.

TABLE 23:2

OXIDATION OF VANADIUM

Reagents	Treatment	Final Color	Indicated V
HNO_3-H_2SO_4	Fumed 5 min	Brown cast	0.7 mg
HNO_3-H_2SO_4	Fumed 1 min	Sl. brn. cast	0.95 mg
HNO_3-H_2SO_4	HNO_3, just boiled	Yellow	1.03 mg
HNO_3-$HClO_4$-H_2SO_4	$HClO_4$, just boiled	Yellow	1.00 mg
H_2O_2-H_2SO_4	Fumed 2 min	Brn.-pink	—

These data are self-explanatory; it is seen that a brown color in the final solution may indicate reduced vanadium. A spectrographic method for vanadium in fuel oils has been described by Anderson and Hughes; [13] an X-ray spectrographic method has been developed by Davis and Hoeck. [14]

PREPARATION

1. REAGENTS:
> *Phosphoric Acid Reagent:* Mix one volume of 85% H_3PO_4 with two volumes of water.
> *Sodium Tungstate Reagent:* Dissolve 16.5 grams of $Na_2WO_4 \cdot 2H_2O$ in 100 ml of water.
> *Standard Vanadium Solution:* See *REAGENTS*, Section 1.

2. STANDARDIZATION:

Transfer 0.5, 1.0, 1.5, 2.0, and 2.5 mg of vanadium to 150-ml beakers, including a blank. To each beaker add 5 ml of H_2SO_4, 1 ml of HNO_3, and 1 ml of $HClO_4$, then heat it on a hot plate until the $HClO_4$ begins to boil. Remove the beakers from the heat, and cool. Dilute to about 50 ml, add 10 ml of H_3PO_4 $(1 + 2)$, and 5 ml of Na_2WO_4 reagent. Heat just to boiling, remove from the heat, and allow to cool to room temperature. Transfer the solutions to 100-ml volumetric flasks, dilute to the mark, and mix thoroughly. Measure the transmittances of the solutions at 430 mμ using 13-mm cuvettes, and plot against the corresponding milligrams of vanadium on semi-logarithmic graph paper.

PROCEDURE

1. To choose an appropriate amount of sample, use the relation:

$$\text{Size of Sample} = 1500/\text{ppm V}$$

Transfer the weighed sample to a 400-ml beaker, and treat it as described in Section 1A.

> NOTE: Samples less than 3 grams should be wet-oxidized directly. If the sample size is less than 50 grams, 5 drops of fuming sulfuric acid are sufficient.

2. When the wet oxidation is complete, and the volume of H_2SO_4 has been reduced to about 5 ml, add 1 ml each of HNO_3 and $HClO_4$, and heat until the $HClO_4$ begins to boil. Remove it from the heat, cool, dilute to 50 ml with water, and boil 5 minutes.

> NOTE: The volume of sulfuric acid is not critical; volumes between 5 and 10 ml yield the same result.

3. If there is an insoluble residue, filter the solution through Whatman No. 40 paper, wash with a little water, and discard the residue. To the filtrate add 10 ml of H_3PO_4 $(1 + 2)$ and 5 ml of Na_2WO_4 reagent, heat it just to boiling, then allow it to cool to room temperature.

NOTE: Excessive boiling or rapid cooling causes the formation of a precipitate. Low results ensue if this is filtered, but fairly satisfactory results are obtained by centrifuging the cuvette before measuring the transmittance.

4. Transfer the cool solution to a 100-ml volumetric flask, dilute to the mark, and mix thoroughly. Measure the transmittance of the solution at 430 mμ, and read the corresponding milligrams of vanadium from the prepared calibration curve.

3. Volumetric Determination of Vanadium

The procedure given here is for the determination of vanadium in synthetic oil-soluble compounds used for research. As concentrations usually range from two to five per cent, a small sample can be weighed, and wet-oxidized directly. The acid solution is then diluted, and poured through a Jones reductor, the effluent being collected under a mixture of phosphoric acid and ferric ammonium sulfate, and titrated with standard permanganate. Vanadium is reduced to the bivalent state in the reductor, and then oxidized to the vanadyl ion in the receiver according to the following equations:

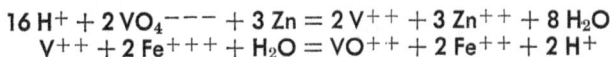

$$16\,H^+ + 2\,VO_4^{---} + 3\,Zn = 2\,V^{++} + 3\,Zn^{++} + 8\,H_2O$$
$$V^{++} + 2\,Fe^{+++} + H_2O = VO^{++} + 2\,Fe^{++} + 2\,H^+$$

The reaction between vanadyl ion and permanganate is slow near the equivalence point, and care must be taken not to under-titrate. When all of the vanadium has been oxidized, an end point is obtained which is stable for one minute or more.

It is presumed that only vanadium is present. Chromium, iron and molybdenum would interfere in the method given here, so if they are present the method of Hamner [15] can often be used to advantage.

PROCEDURE

1. Weigh a 1 to 2 gram sample, and transfer it to a 400-ml beaker. Add 10 ml of H_2SO_4, and oxidize the organic material

as usual with HNO_3 and 30% H_2O_2, completing the oxidation with $HClO_4$. The final volume of H_2SO_4 should be 5 ml or less.

> NOTE: Care must be taken that no nitric acid remains in the solution as it is reduced by zinc to hydroxylamine and other compounds that consume permanganate.

2. Cool the solution, dilute it to approximately 200 ml, and pour it through a Jones reductor, collecting the effluent under a mixture of 25 ml of 10% ferric ammonium sulfate and 5 ml of H_3PO_4. Wash the column with water, and titrate the contents of the receiver with standard $N/10$ $KMnO_4$, continuing the titration until an end point is obtained that persists for one minute. Record the milliliters of permanganate consumed, and correct for a reagent blank (equivalent weight of vanadium here is 16.98).

> NOTE: As the end point is slow, it is advisable to time with a watch the addition of each tenth of a milliliter.

References

1. L. K. BEACH and J. E. SHEWMAKER, "The Nature of Vanadium in Petroleum," *Ind. Eng. Chem.*, 49:1157 (1957).
2. F. M. WRIGHTSON, "Determination of Traces of Iron, Nickel, and Vanadium in Petroleum Oils," *Anal. Chem.*, 21:1543 (1949).
3. FRITZ FEIGL, *Spot Tests, Volume I, Inorganic Applications*, p. 121, Elsevier, New York (1954).
4. J. I. HOFFMAN and G. E. F. LUNDELL, "Volatilization of Metallic Compounds from Solution in Perchloric or Sulfuric Acid," *J. Research Nat'l Bur. Standards*, 22:465 (1939).
5. G. F. SMITH, "The Dualistic Versatile Reaction Properties of Perchloric Acid," *Analyst*, 80:16 (1955).
6. M. C. K. JONES and R. L. HARDY, "Petroleum Ash Components and Their Effect on Refractories," *Ind. Eng. Chem.*, 44:2615 (1952).
7. I. G. SLATER and W. L. PARR, "Marine Boiler Deterioration," *J. Am. Soc. Naval Engrs.*, 62:405 (1950).
8. F. E. CLARKE, "Vanadium Ash Problems in Oil-Fired Boilers," *Ibid.*, 65:253 (1953).
9. O. I. MILNER, J. R. GLASS, J. P. KIRCHNER, and A. N. YURICK, "Determination of Trace Metals in Crudes and Other Petroleum Oils," *Anal. Chem.*, 24:1728 (1952).
10. L. W. GAMBLE and W. H. JONES, "Determination of Trace Metals in Petroleum," *Ibid.*, 27:1456 (1955).
11. Am. Soc. Testing Materials, *ASTM Standards on Petroleum Products and Lubricants*, D1548–58T (1958).
12. E. R. WRIGHT and M. G. MELLON, "The Phosphotungstate Method for Vanadium," *Ind. Eng. Chem., Anal. Ed.*, 9:251 (1937).

13. J. W. ANDERSON and H. K. HUGHES, "Spectroscopic Determination of Vanadium in Residual Fuel Oils," *Anal. Chem.*, 23:1358 (1951).
14. E. N. DAVIS and B. C. HOECK, "X-Ray Spectrographic Method for the Determination of Vanadium and Nickel in Residual Fuels and Charging Stock," *Ibid.*, 27:1880 (1955).
15. H. L. HAMNER, "Determination of Chromium and Vanadium in Steel," *Chem. Met. Eng.*, 17:206 (1917).

Chapter 24

ZINC

The presence of zinc in crude oils is rare, and the element is seldom encountered in refinery stocks. Zinc additives have been used in lubricating oils for many years, however, and zinc naphthenate is an important paint drier. The analyses of these products are covered in Sections 1 and 2 here and two methods for the final determination are described in Sections 3 and 4.

Zinc is one of several metals that attack silver bearings and wrist-pin bushings in diesel locomotives; its concentration in the lubricating oil should not exceed about 20 parts per million. Small concentrations of the element are best determined by spectrographic methods.[1]

Before presenting the detailed procedures several special methods should be mentioned. Gerhardt and Hartmann[2] have applied a titration with Versenate for determining zinc in lubricating oils and concentrates; and for the same purpose Marple et al.[3] described a nonaqueous photometric titration with dithizone. Davis and Van Nordstrand[4] have given a method of spectroscopy with x-ray fluorescence for determining zinc and other metals in lubricating oils.

1. Determination of Zinc in Additives and Naphthenate Driers

Gottsch and Grodman,[5] and Weatherburn et al,[6] have reported that low results are invariably obtained when direct ashing is used for determining zinc in naphthenate driers; this is caused by sublimation of zinc oxide. Because of the tendency of many zinc additives to "creep," slightly low results are often obtained by wet oxidation, and for this

246

reason the decomposition of these products by fusion with potassium pyrosulfate is prescribed in the procedure here. To illustrate the magnitude of this effect samples of an additive were decomposed by wet oxidation, and by pyrosulfate fusion; the final determination of zinc was made by potentiometric ferrocyanide titration, as described in Section 3. The results for zinc were as follows: pyrosulfate fusion, 8.35, 8.37; wet oxidation, 8.15, 8.19.

The presence of sulfur in many additives leads to the formation of sulfide in the fusion mixture. As this interferes in the potentiometric procedure (by reducing ferricyanide) it must be removed; this is conveniently accomplished by cooling the melt, adding a few drops of sulfuric acid, and again fusing the mixture. Even if the sample does not contain sulfur this step should not be omitted because there may be some reduction of the pyrosulfate itself by the organic material in those samples. If the gravimetric procedure in Section 4 is used to finish the determination a double precipitation must be made; normally the volumetric procedure in Section 3 is most convenient and rapid.

PROCEDURE

1. Place approximately 5 grams of fused potassium pyrosulfate in a Coors No. 2 porcelain crucible. Melt the flux over a burner, and allow it to cool while rotating the crucible in such a way that its interior is uniformly coated to approximately one-third its height.

NOTE: If many determinations are to be made several crucibles may be prepared in advance.

2. By means of a Lunge weighing bottle accurately weigh a 1-gram sample, and transfer it to the prepared crucible. Add about 10 grams of potassium pyrosulfate, covering the sample loosely, and heat it until the sample ignites. Apply heat as necessary to maintain combustion, and continue heating carefully until only a small amount of carbon remains. Cover the crucible, and heat strongly with a blast burner until all carbon is destroyed.

NOTES: Naphthenate driers must be weighed in Lunge weighing bottles; lubricating oil additives are conveniently handled in ¾-oz soufflé cups.

Sometimes the flux solidifies while much carbon still remains. If this happens, cool the flux, and add about 6 drops of sulfuric acid. Upon heating the flux will again liquefy, but care must be taken to minimize spattering. If too much sulfuric acid is used the mixture will boil and spatter violently.

3. Allow the melt to cool, remove the cover, add 2 drops of H_2SO_4, heat it again until the mixture liquefies, then cool it to room temperature. To the crucible add 20 ml of 1:4 H_2SO_4, warm it on a hot plate until the salt dissolves, then wash the contents of the crucible into a 400-ml beaker. Police the cover and crucible with a stirring rod, also apply a stream of water from a wash bottle, and transfer any remaining particles to the beaker. Dilute the solution to about 200 ml, heating if necessary to dissolve all salts, and treat the solution by the procedure in Section 3.

NOTES: If sulfuric acid was added, as mentioned in the preceding note, it need not be added here.

The dilute sulfuric acid dissolves the melt much more rapidly than water alone.

Small amounts of silica, barium sulfate, etc., are without effect; the solution need not be filtered.

2. Determination of Zinc in New and Used Lubricating Oils

The concentration of zinc used as compounding in lubricating oils is usually less than 0.1 per cent, but larger amounts are often encountered in used oils. The element can usually be determined in new oils without making any separations if only the common additive metals are present (calcium, barium, phosphorus, and sulfur). In examining used oils, however, account must be taken of additional elements that may be present including aluminum, copper, iron, lead, molybdenum, and silicon.

In the following procedure a sample is prepared by the usual method of soft ashing and wet oxidation, and if the sample is a new oil, no separations need be made. When handling used oils, however, the oxidized sample is diluted and filtered to remove barium, lead, and silica. The filtrate is then treated with cupferron, in order to precipitate iron, molybdenum, and most of any copper present. After filtering again, the remainder of the copper is removed from the filtrate by treating with hydrogen sulfide. Left in the filtrate are aluminum, calcium, phosphorus, and zinc. This filtrate is evaporated to fumes of sulfuric acid, and excess cupferron is destroyed with nitric acid. The resulting sulfuric acid solution is diluted, and zinc is determined by the procedure in Section 3.

Studies by Waldbauer et al.[7] have shown that zinc is coprecipitated by lead sulfate. With the amounts of lead encountered in lubricating oils, however, this effect is not usually significant. For example, 300 mg of lead sulfate carries down about one milligram of zinc; the coprecipitation is attributed to formation of mixed crystals.

PREPARATION

1. REAGENT:

> *Cupferron Solution:* Dissolve 6 grams of cupferron, Eastman No. 82, in 100 ml of water. The solution is not stable.

PROCEDURE

1. Weigh a suitable sample, transfer it to a 400-ml beaker, warm it on the steam plate, and add 5 drops of fuming H_2SO_4. Stir thoroughly, wipe the rod with a piece of filter paper, and drop the paper into the beaker. Place the beaker in a muffle can, ignite the oil with a flame, and allow the sample to burn freely until only a soft ash remains. Burn the loose carbon from the upper walls of the beaker with a blast burner, avoiding direct heating of the bottom, and then allow the beaker to cool.

> NOTES: For most lubricating oils a 20–30 gram sample is satisfactory.
> If the sample is a new oil, or if the absence of lead is certain, the treatment with fuming sulfuric acid may be omitted.

2. Add about 10 ml of H_2SO_4, heat to fumes, and oxidize the carbonaceous material with HNO_3 and 30% H_2O_2 as usual, finishing with $HClO_4$. Reduce the volume of H_2SO_4 to about 5 ml, and cool. Add 100 ml of water, heat to boil-

ing, and boil 5 minutes. Cool the solution in an ice bath, for about 30 minutes, and filter it through Whatman No. 42 paper.

> NOTES: Insoluble material may include silica, barium, calcium, or lead sulfates; the residue should be washed with 1:100 sulfuric acid. If the solution is clear after boiling the filtration may be omitted.
>
> If the sample is a new lubricating oil treat this solution by the procedure in Section 3. If it is a used oil proceed as follows:

3. Dilute filtrate from step 2 to 200 ml, chill it in an ice bath for a few minutes, and add in drops and with vigorous stirring a cold 6% aqueous solution of cupferron until precipitation is complete. This stage may be judged by allowing the precipitate to settle, and adding a drop of cupferron solution to the supernatant liquid. A white flash of salt that immediately dissolves indicates that precipitation is complete.

Care must be taken not to mistake the white precipitate of molybdenum cupferrate, which does not dissolve. In the presence of a large precipitate this test may fail, and the first portion of filtrate should always be tested with a few drops of reagent. A clear supernatant liquid without any milky appearance is obtained when the reagent is in excess.

> NOTE: If the reagent is added too rapidly the precipitate will separate in large gummy masses that are difficult to wash.

4. Filter the precipitate on Whatman No. 40 paper, testing the first few milliliters of filtrate for completeness of precipitation, and wash the precipitate with four or five 10-ml portions of a solution containing 2 ml of cupferron reagent in 100 ml of 5% H_2SO_4. Discard the precipitate.

> NOTE: An acid-cupferron solution is used for the washing to prevent possible precipitation of cupferrates that separate at lower acidities, and to minimize the solubility of the precipitate.

5. To the filtrate add 10 ml of HNO_3, evaporate to fumes of H_2SO_4, and oxidize the remaining organic material as usual, finishing with $HClO_4$. The final volume of H_2SO_4 should be 3–5 ml. Cool, dilute to 200 ml with water, and boil

for 5 minutes. Cool the solution, and treat it by the procedure in Section 3.

NOTE: Ordinarily this solution is clear after boiling; if relatively large amounts of calcium are present some calcium sulfate may fail to dissolve.

3. Potentiometric Determination of Zinc

In this section the potentiometric titration of zinc with standard potassium ferrocyanide using the ferri-ferrocyanide electrode is briefly described. As Kolthoff and Furman [8] have given a complete discussion of this method, only a few details are mentioned here.

The reaction is represented by the following equation:

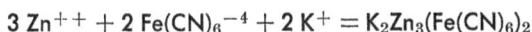

$$3 \, Zn^{++} + 2 \, Fe(CN)_6{}^{-4} + 2 \, K^+ = K_2Zn_3(Fe(CN)_6)_2$$

The titration is carried out in a hot solution at about pH 2. Among the interfering elements the most troublesome are copper and iron; the effect of the latter is avoided by the addition of fluoride ion. Both of these elements are separated in Section 2. As the presence of ammonium sulfate increases the error in hot solution, excess sulfuric acid is neutralized with potassium hydroxide; the potassium sulfate thus formed improves the titration. Large amounts of aluminum render the electrode insensitive, but the amounts normally encountered in petroleum products are without effect. The standard ferrocyanide solution is stabilized by addition of a small amount of sodium carbonate, but its titer should be checked occasionally.

PREPARATION

1. REAGENTS:

Potassium Ferricyanide Solution: Dissolve 10 grams of $K_3Fe(CN)_6$ in 1000 ml of water.

Potassium Ferrocyanide Solution: Dissolve 10 grams of $K_4Fe(CN)_6 \cdot 3H_2O$, and 2 grams of Na_2CO_3 in 1000 ml of water, mix well, and store in a brown bottle. Standardize this solution against the standard zinc solution by carrying a 50-ml aliquot of the latter through the procedure that follows.

$$\text{titer, mg Zn}^{++}/\text{ml} = \frac{\text{mg of zinc in aliquot}}{\text{ml of ferrocyanide to titrate}}$$

Potassium Hydroxide Solution: Dissolve 300 grams of KOH in 1000 ml of water, cool, and mix.

Zinc Standard Solution: Dissolve 1.25 grams of ZnO in a slight excess of sulfuric acid, and dilute to 1000 ml. Using a 50-ml aliquot, standardize the solution by the gravimetric procedure in Section 4.

NOTE: If desired, a 1-gram portion of metallic zinc may be dissolved, diluted to volume, and used without standardization.

PROCEDURE:

1. To the solution prepared by the procedure in Section 1 or 2 (200-ml volume) add 1 ml of 2,4-dinitrophenol indicator, and neutralize by addition of 30% KOH in drops until the indicator just turns yellow; then add 1–2 drops of H_2SO_4. Heat the solution to about 80°C on a hot plate, then transfer it to a titration stand equipped with a hot stage, a platinum electrode, and a fiber-type calomel electrode with a salt bridge. Any commercial *p*H meter or potentiometer may be used to follow the titration.

NOTES: This neutralization produces a *p*H of about 2.

Hot stages are available from most laboratory supply houses; it is necessary to keep the solution hot during the course of the titration.

2. To the solution add 1 gram of NaF, and 5 ml of 1% potassium ferricyanide solution. Using a millivolt meter the initial potential will be from 650 to 800 mv depending upon the concentration of zinc. Titrate slowly until the potential falls to about 600 mv, then add one-tenth milliliter increments at 1 minute intervals, recording the corresponding potentials. With significant amounts of zinc the maximum change in potential is of the order of 90–100 mv.

NOTES: Sodium fluoride is added as a precaution to mask the effect of any ferric ion that may be present. The solution should be discarded without delay after the titration to avoid excessive attack on the beaker.

The ferricyanide solution should not be added until ready to titrate as it is slowly hydrolyzed in the hot solution to ferric ion and hydrogen cyanide. A dark blue precipitate upon addition of ferrocyanide indicates that the determination is worthless.

Equilibrium is reached slowly near the end point, and care must be taken not to over-titrate.

3. Compute the exact titration by the second derivative method (see Chapter 11, Section 5), and calculate the percentage of zinc in the sample, using the ferrocyanide titer determined by the foregoing procedure.

4. Gravimetric Determination of Zinc

This section is included to provide a method of preparing a standard zinc solution, and is not very well suited for determining zinc in the solutions prepared in Section 1 or 2. It gives satisfactory results, however, if a double precipitation is made; it should not be applied to used oils handled by the procedure in Section 2.

PROCEDURE

1. Transfer a 50-ml aliquot (about 50 mg of zinc) of the zinc standard solution prepared in Section 3 to a 400-ml beaker, add 5 grams of NH_4Cl, dilute to 200-ml, and heat to boiling. Slowly add 20 ml of 10% $(NH_4)_2HPO_4$ solution from a pipette, and digest all on the steam plate until the precipitate is granular.

2. Remove beaker from the heat, allow it to stand at least 2 hours, and filter the precipitate on a tared Selas crucible. Wash the precipitate with several 10-ml portions of a hot 1% solution of $(NH_4)_2HPO_4$, and finally with 50% alcohol. Dry the crucible and precipitate in a radiator, then ignite at 900°C for about 45 minutes. Cool it and reweigh, calculating the strength of the solution in mg Zn/ml; (gravimetric factor for zinc from $Zn_2P_2O_7$ is 0.4290).

NOTES: The solution should be just acid to methyl orange. If excess acid is present it must be neutralized with ammonium hydroxide.

If a double precipitation is to be made filter on Whatman No. 40 paper, and redissolve in hydrochloric acid.

References

1. L. E. CALKINS and M. M. WHITE, "Analyze Additive Lubricants in Minutes Instead of Hours with Spectrographic Method," *Nat'l. Petroleum News*, 38, No. 27:519 (1948).
2. P. B. GERHARDT and E. R. HARTMANN, "Determination of Calcium or Zinc Additives in Lubricating Oils and Concentrates," *Anal. Chem.*, 29:1223 (1957).
3. T. L. MARPLE, G. MATSUYAMA, and L. W. BURDETT, "Nonaqueous Titration of Zinc," *Ibid.*, 30:937 (1958).
4. E. N. DAVIS and R. A. VAN NORDSTRAND, "Determination of Barium, Calcium, and Zinc in Lubricating Oils," *Ibid.*, 26:973 (1954).
5. F. GOTTSCH and B. GRODMAN, "An Extraction Method for the Determination of Metals in Boiled Linseed Oil and Driers," *Proceedings, Am. Soc. Testing Materials*, 40:1206 (1940).
6. A. S. WEATHERBURN, M. W. WEATHERBURN, and C. H. BAYLEY, "Determination of Copper and Zinc in Their Naphthenates," *Ind. Eng. Chem., Anal. Ed.*, 16:703 (1944).
7. L. WALDBAUER, F. W. ROLF, and H. A. FREDIANI, "Spectrographic Studies of Coprecipitation," *Ibid.*, 13:888 (1941).
8. I. M. KOLTHOFF and N. H. FURMAN, *Potentiometric Titrations*, 2nd Edition, John Wiley and Sons, New York (1931).

Appendix

THE WICKBOLD APPARATUS FOR COMBUSTION

Much of the information given here, on the Wickbold apparatus for combustion with an oxyhydrogen flame, has been taken from the three papers by R. Wickbold, cited in Chapter 2, Section 9. The complete apparatus is manufactured by Heraeus Quarzschmelze, Hanau, Germany, and it is available in the United States through their American representative.* The application of this method of combustion for the determination of halogens and sulfur is discussed in Chapters 11 and 22, respectively.

A. THE WICKBOLD APPARATUS

The apparatus is shown in operation in Figure 2. In Figure 3 the flow of gases through the apparatus is shown schematically.

The apparatus consists of a probe burner which fits into a water-cooled quartz combustion chamber connected to an absorber. Suction applied to a spray trap on the top of the absorber draws the combustion gases through a suitable absorbing solution. An oxygen-hydrogen pilot flame ignites the sample which is then oxidized completely in an excess of oxygen. A capacious receiver is required to accommodate the large volume of water formed by the combustion of hydrogen. As the temperature of the flame exceeds 2,000°C, the combustion chamber must be cooled by a rapid flow of water to prevent its being melted, and to minimize attack

---

* Engelhard Industries, Inc., Amersil Quartz Division, 685 Ramsey Avenue, Hillside, New Jersey.

of the quartz by traces of alkali metals which are always present.

The apparatus is regulated by four stopcocks which are seen on the raised portion at the front of the base in Figure 2; their connections are shown schematically in Figure 3. The functions of the stopcocks are as follows:

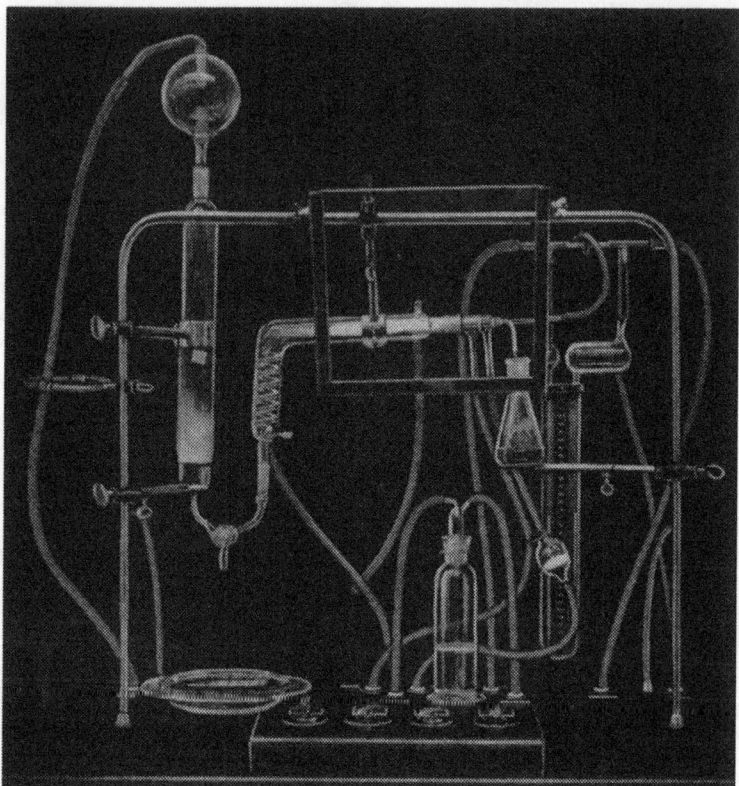

FIG. 2. WICKBOLD APPARATUS FOR OXYHYDROGEN COMBUSTION

1. SELECTOR STOPCOCK

This is a three-way stopcock, only two positions of which are used. There is, or should be, a black dot on one end of the key which shows the stopcock position.

a) Suction position: The air flow under vacuum is adjusted by setting the selector cock (1) with the black dot over the word "suction" on the base plate. With the oxygen cock (2) opened, the hydrogen cock (3) and vacuum cock (4) closed, and the vacuum pump turned on, suction is applied to the vacuum cock which is connected with the top of the spray trap. When the vacuum cock (4) is opened, air is drawn through the absorber, burner, oxygen line, open oxygen cock (2), and flowmeter through the selector cock (1), which in "suction" position is open to the atmosphere. The flow is regulated with the vacuum cock (4), to produce a difference of 22–24 mm of mercury on the flowmeter.

   b) *Pressure position:* When the selector cock (1) is moved one-quarter turn counterclockwise the oxygen supply line is connected to the flowmeter. The outlet of the flowmeter is attached to a Y-tube; one leg of this is attached to the capillary tube on the rear of the burner, and the other leads to the oxygen cock (2) through which most of the oxygen is supplied to the burner. With the burner disconnected from the combustion chamber, the flow is regulated by opening the oxygen cock to produce a difference of 16–18 mm of mercury on the flowmeter.

2. OXYGEN STOPCOCK

   As can be seen in Figure 3, this cock regulates the flow of oxygen to the burner. The oxygen flow is initially set a few mm of mercury less than the vacuum flow. This results in a slight vacuum in the combustion chamber which draws the sample up into the burner. The oxygen cock (2) controls the flow only in the large line to the bottom of the burner. The oxygen supply to the center tube of the burner is restricted by a capillary and therefore, if the oxygen cock (2) is turned toward the "off" position after the initial regulating, the vacuum in the combustion chamber increases, and the sample is drawn in more quickly. The sample flow and the rate of burning are thus controlled entirely by the oxygen cock (2).

   It should be noted that when the selector cock (1) is in the "pressure" position the flowmeter indicates only the oxygen flow. As the flow of hydrogen also contributes to the pressure in the combustion chamber, care must be taken not to open the oxygen cock (2) too far when regulating the flame. If the flow is increased too much a positive pressure will be developed in the combustion chamber which will blow the burner backward out of the sleeve. The danger is increased by the relatively high pressure (15 psi) behind the stopcock, which makes the regulating very sensitive.

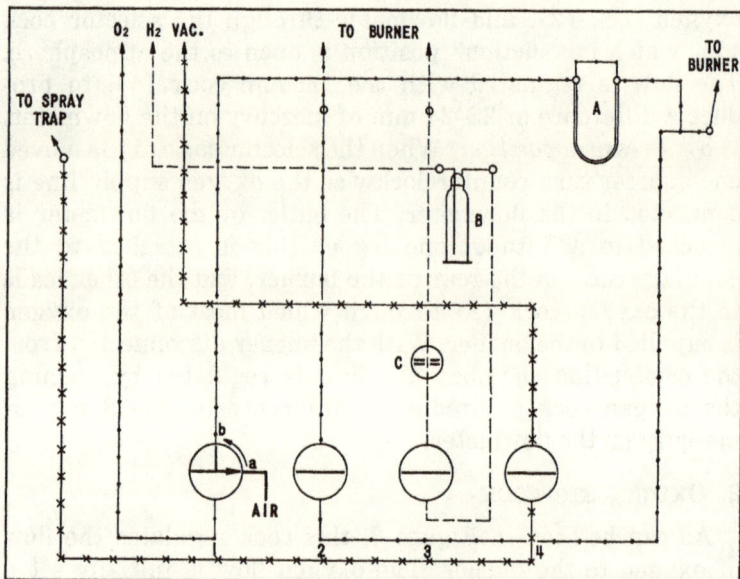

FIG. 3. DIAGRAM OF WICKBOLD APPARATUS
a) Flow meter; *b*) Hydrogen bubbler; *c*) Flame arrester;
1) Selector; 2) Oxygen; 3) Hydrogen; 4) Vacuum.

3. HYDROGEN STOPCOCK

Figure 3 shows that the hydrogen passes by way of a bubble counter, through the hydrogen cock (3) directly to the burner. The length of the flame is adjusted initially to 1–2 cm with the hydrogen cock (3); the setting is not changed during the combustion of the sample. Figure 1 shows a flame arrester in the hydrogen line to the burner, consisting of a coarse fritted-glass plate sealed in a bulb. This is not included with the apparatus, but it is an important safety feature; it need not be as large as the one shown in the photograph.

4. Vacuum stopcock

The function of this stopcock was covered in describing the operation of the selector cock (1). Its position is not changed after the initial setting of the flow.

B. AUXILIARY EQUIPMENT

The requirements for operating the Wickbold apparatus assume available sources of hydrogen, oxygen, vacuum, and water. It is good safety practice to stand hydrogen cylinders outside the laboratory, and for convenience the oxygen cylinder should be in the same place. As considerable oxygen is required accessibility is important; the hydrogen consumption is small. The cylinders should be equipped with station service regulators preset at about 50 psi. These are connected by copper tubing (enclosed in electrical conduit for protection) with low pressure regulators mounted outside a window in the laboratory. Copper tubing from the regulators enters the laboratory through slots in the window sash and is connected to the apparatus with rubber tubing secured by hose clamps. A small block valve in each line inside the window is convenient for standby and emergency shutoffs.

A vacuum pump with a capacity of about 2,000 liters per hour is required. The amount of gas transported is less than this, but the resistance of the absorbing solution and fritted plate must be overcome. Oil pumps should not be used as they would be badly corroded by the water-saturated gas stream of carbon dioxide and excess oxygen, and the passage of oxygen through the warm oil is also a hazard. A suitable source of vacuum is an Eberbach air-cooled rotary air blast and suction apparatus. When operated by a one-quarter horsepower motor at 1750 rpm its capacity is about 2550 liters per hour. These pumps, which are stocked by most laboratory supply houses, are inexpensive, compact, and easily replaced. As leakage through the three-way stopcock on the bottom of the absorber can decrease the vacuum somewhat, this stopcock, as well as the vacuum control cock (4), should be lubricated with a high-vacuum grease and secured with a tension clip.

A strong flow of water is required for cooling the combustion chamber, and the rubber tubing connections should be secured with hose clamps.

C. OPERATION OF WICKBOLD APPARATUS

The apparatus is assembled initially as described in the manufacturer's instructions, the cooling water and gas connections being made according to the name plates on the base of the apparatus. The manometer (flowmeter) may be filled with water, or with mercury; the following instructions refer to the latter. The hydrogen bubble counter is filled to the mark with water, and the stopper secured with a spring arrangement. A flame arrester should be placed in the hydrogen line to the burner, as previously mentioned. The cooling water is turned on full, and the operation should proceed as follows:

1. Set the three-way stopcock on the absorber so that all outlets are closed.

2. Charge the absorber with 50 ml of a suitable absorbing solution (see chapters on halogens and sulfur, and Section D), and attach the spray trap.

3. Insert the probe burner in the combustion chamber.

4. Close the hydrogen and vacuum cocks, and open the oxygen cock wide.

5. Turn the vacuum pump on, and set the selector cock to the "suction" position.

6. Open the three-way cock on the absorber so that it is connected to the rest of the apparatus.

7. Open the vacuum cock slowly until a difference of 22–24 mm of mercury is produced on the manometer.

8. Disconnect the suction burner from the combustion chamber.

9. Set the selector cock to the "pressure" position, and close the oxygen cock.

10. Open the reducing valve on the oxygen supply line to about 15 psi. (If a block valve is included after the regulator, as recommended in Section A, it should also be opened.)

11. Open the oxygen cock until a difference of 16–18 mm of mercury is produced on the manometer.

12. Open the reducing valve on the hydrogen supply line to about 5 psi. (Also the block valve, if provided.)

13. Open the hydrogen cock until a flow of 2–4 bubbles per second is obtained through the bubble counter.

14. Light the burner with a pilot flame, and adjust the length of the oxyhydrogen flame to about 1 cm with the hydrogen cock.

15. Replace the burner in the combustion chamber, taking care that the flame does not impinge on the polished quartz surfaces.

16. Place a flask containing the weighed sample under the suction capillary tube.

17. When the liquid reaches the tip of the burner, adjust the flame to fill the combustion chamber from one-half to three-fourths of its length, using the oxygen cock.

> NOTE: Turning the oxygen cock towards the "off" position *increases* the sample flow and length of the flame. (See Section A.) Make adjustments *slowly*, as the regulation is very sensitive.

18. After the sample is burned the sample container is rinsed with some pure solvent which is also aspirated through the burner. Acetone, methanol, methyl isobutyl ketone are suitable.

19. Remove the empty sample container, turn off the hydrogen, and detach the burner.

20. Turn the selector cock to "suction," turn the oxygen and vacuum cocks off, turn the three-way stopcock on the absorber to close all outlets, and switch off the vacuum pump. If inside block valves are available they should be closed to prevent the escape of gas should a stopcock be unseated.

21. Drain the absorbing solution, and rinse the apparatus, taking care that the fritted plate and portion of the absorber below it are washed thoroughly. The handling of the solution is covered in the chapters on halogens and sulfur.

D. NOTES ON PROCEDURE

1. When gaining experience in regulating the flame and feed rate, or when a sample with unknown burning characteristics is to be burned, it is advisable to make an initial

adjustment with an oxygen-containing solvent such as methyl isobutyl ketone. Such compounds are not likely to smoke, and after the feed rate and flame length have been regulated, the flask containing the sample can be placed under the suction capillary, with only minor adjustment necessary.

2. In lighting the oxyhydrogen pilot flame care must be taken not to open the hydrogen cock so far that the top of the bubble counter is lifted, for if this occurs there is likely to be an explosion.

3. If too little oxygen is used in the initial regulating there may be flooding during the burning of alcohols, ketones, and the like; heavier materials and aromatics will probably smoke. At incipient smoking the tip of the flame becomes yellow, or reddish, and if this is observed, the flow of oxygen should be increased immediately (flow of sample decreased).

As aromatics burn with an intense white flame, colored glasses are required to protect the eyes.

4. If it becomes necessary to stop burning during the course of a combustion, the vessel containing the sample must be removed from the sample tube before the hydrogen is turned off, or the liquid will continue to flow and flood the combustion chamber.

5. Samples with high vapor pressures may not feed evenly and the back-and-forth surging can cause a carbon deposit in the sample outlet at the tip of the burner which eventually plugs the line completely. This deposit can be removed by applying air pressure to the sample inlet, and holding the tip of the burner in an oxygen-gas flame; this should be done by a glass-blower.

6. The surging of samples of high vapor pressure can be prevented by chilling the flask containing the sample, or by burning rapidly. If the sample is fed slowly, it is vaporized as the stream reaches the warm region near the tip of the burner, and the vapor pushes the liquid column back down the capillary tube. Burning then ceases until the liquid is again pulled up the tube into the pilot flame. In a few seconds the process is repeated. By feeding rapidly, burning at such a rate that the entire combustion chamber is filled,

the liquid moves through the hot zone in the vicinity of the tip of the burner too rapidly to vaporize, and the sample then burns smoothly. Diluting with a suitable solvent may also control surging.

7. When burning heavy materials, such as cable oils, and similar viscous materials that require dilution, it is desirable to use as heavy a solvent as possible (e.g., a highly refined kerosene). When light solvents are used as diluents, the heavier material is deposited on the burner tip, and in the combustion tube. In some instances plugging occurs, as mentioned in Note 5.

8. Deposits of carbon and tar in the combustion chamber can be removed by closing the three-way stopcock, filling the condenser and chamber with acetone, and brushing the interior with a test tube brush. The acetone is drained off and the assembly dried by suction. If extensive smoking occurs, the soot must be removed with cleaning solution.

9. When shutting the equipment down at night, the gas supply regulators should be turned off and the pressure on the stopcocks relieved by opening them briefly. The inlet tube to the bubble counter should be closed with a pinch clamp. If this is neglected, temperature changes and diffusion cause water to be drawn back into the hydrogen supply line. Water in this line causes momentary interruptions in flow as evidenced by flashback to the flame arrester. This in turn causes the sample to flood the combustion chamber, extinguishing the flame, or smoking the chamber. When starting operations, the pinch clamp should be removed after step 12 in the sequence of operations in Section C.

10. If the equipment is operated continuously during the working day, an entire cylinder of oxygen may be consumed. A cylinder of hydrogen lasts several weeks.

E. BEHAVIOR OF THE ELEMENTS
IN THE COMBUSTION

The Wickbold combustion method is best suited for the determination of relatively small concentrations of halogens (see Chapter 11) and sulfur (see Chapter 22) in petroleum products. The velocity of the flow through the absorber is so great that absorption may be incomplete when concentrated samples are burned, and the results will be erratic. It is usually possible, however, to absorb these elements completely by diluting the sample with a suitable solvent.

As the oxyhydrogen flame operates at a temperature above 2,000°C, all carbon-fluorine bonds are ruptured, with the production of hydrogen fluoride. The velocity of the gas through the apparatus is too great to permit any reaction with the quartz combustion chamber, and the evolved hydrogen fluoride is quantitatively recovered in dilute sodium hydroxide. Small amounts (1–5 mg) are determined by titration with standard thorium nitrate; larger amounts may be precipitated as lead chlorofluoride, and determined volumetrically with standard silver nitrate.

Chlorine compounds are decomposed with the evolution of hydrogen chloride, and if the content is high, also traces of free chlorine. Water is a suitable absorbent for samples containing 100 ppm, or less. If more concentrated samples are burned, dilute sodium hydroxide containing a few drops of hydrogen peroxide should be used. It is usually possible to burn a large enough sample to permit a potentiometric determination of chloride, although Wickbold describes a nephelometric procedure for small concentrations. (Chapter 2, Reference 30.)

Bromine compounds yield a mixture of hydrogen bromide and free bromine; dilute alkali and peroxide recover the element as bromide.

Iodine compounds yield a mixture of iodine pentoxide and free iodine, the former being deposited as a white solid in the combustion chamber. This is washed out with water, the washings being sucked into the absorber which is charged with dilute alkali and thiosulfate. As iodine is not ordinarily encountered in petroleum, it will not be considered here. The original article by Wickbold (Chapter 2, Reference 27) should be consulted for details.

Sulfur is evolved from the burner as sulfur dioxide which is oxidized by peroxide in the absorber to sulfuric acid. The ubiquity of nitrogen, which is collected in the absorber as nitric acid, precludes an acidimetric determination of sulfur. Wickbold (Chapter 2, Reference 28) applied a photometric titration, the conditions of which require that the absorbing solution contain only hydrogen peroxide. Using water and hydrogen peroxide as the absorbing solution, he found that recovery was quantitative in the 1–100 ppm range, with a precision of ± 2 per cent. The useful range of the method is 100–600 micrograms of sulfur.

Traces of alkali metals cause gradual deterioration of the tip of the burner so samples containing significant amounts of these elements should not be burned in the apparatus. As their attack on the walls of the combustion chamber is minimized by the cooling water, only a little damage is apparent after many combustions.

Heavy metals deposit as oxides on the tip of the burner. They can be dissolved in nitric, or hydrochloric acid, but samples containing them should not be burned.

The effect of phosphorus is very serious. As phosphorus pentoxide is deposited on the tip of the burner which operates at red heat, this forms a low-melting eutectic with the quartz, which rapidly destroys the burner. Samples that contain phosphorus, therefore, must never be burned in the Wickbold equipment.

INDEX

267